Water Recycling and Low-Carbon Development

In Focus – Special Book Series

Water Recycling and Low-Carbon Development

Editors

Yun-Xiao Chong, Tian-Yin Huang, Feng-Min Li, Guang-Li Liu and Qian-Yuan Wu

IWA PUBLISHING

中建环能 SCIMEE

Published by **IWA Publishing**
Unit 104–105, Export Building
1 Clove Crescent
London E14 2BA, UK
Telephone: +44 (0)20 7654 5500
Fax: +44 (0)20 7654 5555
Email: publications@iwap.co.uk
Web: www.iwapublishing.com

First published 2022

British Library Cataloguing in Publication Data
A CIP catalogue record for this book is available from the British Library

ISBN: 9781789063073 (paperback)

Contents

© 2022 The Authors

Water Recycling and Low-Carbon Development doi: 10.2166/wrd.2021.002

Editorial: Water recycling and low-carbon development

Water treatment and reuse is the key strategy to address the problems of water shortages. To facilitate communication and application of technology and research in water reclamation, the National Conference on Water Treatment and Reuse of China was held annually from 2017 by the Committee of water treatment and reuse, Chinese society for environmental sciences.

The theme of the 5th conference was water recycling and low-carbon development. The conference, held from April 16th to 18th in 2021, brought together researchers working in different areas related to water treatment and reuse from China, USA, Australia, Ireland and many other countries and regions. More than 1,200 researchers from universities, institutes and companies attended the conference.

The conference covered 20 main topics:

1. Physical separation theory and technologies
2. Chemical transformation theory and technologies
3. Biological purification theory and technologies
4. Natural treatment theory and technologies
5. Water treatment theory and technology based on microalgae
6. Chemical risk and control
7. Disinfection and microbial risk control
8. Standard and risk control of reclaimed water environment reuse
9. Reclaimed water storage and reuse
10. Water quality detection and water characteristic research methods
11. Standards and policy for water quality control and water resource management
12. New materials, reagents, and equipment for water treatment
13. Design and intelligent operation of water treatment process
14. Urban water system and water recycling
15. Industrial water system and water recycling
16. Rural water system and water recycling
17. Pollution control and utilization of rainwater runoff
18. Water environment constructing and regional water recycling
19. Watershed environment remediation and water recycling
20. Desalination technology and processes

In this special issue of *Water Reuse*, 10 papers that focus on cutting-edge technology and system for water treatment and reuse, were selected among papers presented at the 5th National Conference on Water Treatment and Reuse of China. This special issue is sponsored by CSCEC Scimee Sci. & Tech. Co., Ltd to cover the article processing charges.

Each of these 10 papers focuses on a certain link of water reclamation and reuse, together forming the entire chain of treatment-distribution-detection-utilization. The first six articles focus on water treatment and risk control. Zhou *et al.* (2021) investigate the performance and mechanism of organic pollutants degradation in a photoelectrochemical system. Yin *et al.* (2021) study the removal efficiencies and mechanism of atrazine degradation by peroxymonosulfate activated with $CuCo_2O_4$ nano-particles. Hu *et al.* (2021) present a study on diatrizoate degradation by ozone and the formation of disinfection byproducts in the sequential chlorination. Yi *et al.* (2021) investigate the effect of commercial probiotics immobilized by various carriers on the remediation of aquaculture water and sediment. Yu *et al.* (2021) design a powder-activated carbon pre-coating membrane unit as a pretreatment for ultrafiltration system to alleviate membrane fouling and improve performance. Ren *et al.* (2021) investigated the efficiency of micro-flocculation as a pretreatment approach in alleviating ultrafiltration membrane fouling caused by organic matters in treated wastewater.

Further, two articles are related to distribution and water quality monitoring. Xu *et al.* (2021) report the effects of five pipe materials on water quality deterioration of reclaimed water produced by reverse osmosis process in large pilot-scale distribution systems by long-term running tests. Mao *et al.* (2021) propose a potential rapid and simple method for bacterial detection in water by excitation-emission matrix (EEM) fluorescence spectroscopy.

Lastly, two articles provide new insights to water recycling and watershed environment remediation. Liu *et al.* (2021) study the effects of mixed irrigation using brackish water and reclaimed water on a soil-crop system. Jiang *et al.* (2021) analyze the effects of large-scale compound ecological engineering system on pollution control of the estuary of a lake.

This special issue deals with various problems and investigates different treatment technologies concerning water reuse. In response to the challenges of severe water shortages and the deterioration of water quality, the focus of cutting-edge technology and system for water treatment and reuse is fast-growing worldwide. Progress is needed to provide high quality reclaimed water, minimize the risks to environment and human health, and support use of non-conventional water resources.

Guest Editors

Yun-Xiao Chong

College of Natural Resources and Environment, South China Agricultural University, Guangdong, China

Tian-Yin Huang

School of Environmental Science and Engineering, Suzhou University of Science and Technology, Jiangsu, China

Feng-Min Li IWA

College of Environmental Science and Engineering, Ocean University of China, Shandong, China

Guang-Li Liu

School of Environmental Science and Engineering, Sun Yat-Sen University, Guangdong, China

Qian-Yuan Wu IWA

Tsinghua Shenzhen International Graduate School, Tsinghua University, Guangdong, China.

REFERENCES

Hu, C.-Y., Ren, S.-C., Lin, Y.-L., Zhang, J.-C., Zhu, Y.-Y., Xiong, C. & Wang, Q.-B. 2021 Kinetics of diatrizoate degradation by ozone and the formation of disinfection by-products in the sequential chlorination. *Water Reuse* **11** (4), 560–571.

Jiang, Y., Xu, X., Wu, W., Wang, B., Sun, Y., Chen, X., Shen, Z. & Huang, T. 2021 Analysis on the effect of large-scale compound ecological engineering system on pollution control of the estuary of a lake. *Water Reuse* **11** (4), 649–659.

Liu, C., Cui, B., Hu, C., Wu, H. & Gao, F. 2021 Effects of mixed irrigation using brackish water with different salinities and reclaimed water on a soil-crop system. *Water Reuse* **11** (4), 632–648.

Mao, Y., Chen, X.-W., Chen, Z., Chen, G.-Q., Lu, Y., Wu, Y.-H., & Hu, H-Y. 2021 Characterization of bacterial fluorescence: insight into rapid detection of bacteria in water. *Water Reuse* **11** (4), 621–631.

Ren, L., Liu, C., Meng, T. & Sun, Y. 2021 Effects of micro-flocculation pretreatment on the ultrafiltration membrane fouling caused by different dissolved organic matters in treated wastewater. *Water Reuse* **11** (4), 597–609.

Xu, X., Cui, Y., Wang, Z., Zhang, H., Li, C. & Yu, K. 2021 Water quality deterioration of reclaimed water produced by reverse osmosis process in large pilot-scale distribution systems of different pipe materials. *Water Reuse* **11** (4), 610–620.

Yi, M., Wang, C., Wang, H., Zhu, X., Liu, Z., Gao, F., Ke, X., Cao, J., Wang, M., Liu, Y. & Lu, M. 2021 The *in situ* remediation of aquaculture water and sediment by commercial probiotics immobilized on different carriers. *Water Reuse* **11** (4), 572–585.

Yin, H., Li, J., Yan, H., Cai, H., Wan, Y., Yao, G., Guo, Y. & Lai, B. 2021 Activation of peroxymonosulfate by $CuCo_2O_4$ nano-particles towards long-lasting removal of atrazine. *Water Reuse* **11** (4), 542–559.

Yu, T., Yin, H., Cheng, L. & Bi, X. 2021 Effect of powder-activated carbon pre-coating membrane on the performance of the UF system for wastewater reclamation: a pilot-scale study. *Water Reuse* **11** (4), 586–596.

Zhou, Y., Zhang, G. & Zou, J. 2021 Photoelectrocatalytic generation of miscellaneous oxygen-based radicals towards cooperative degradation of multiple organic pollutants in water. *Water Reuse* **11** (4), 531–541.

Water Recycling and Low-Carbon Development doi: 10.2166/wrd.2021.018

Photoelectrocatalytic generation of miscellaneous oxygen-based radicals towards cooperative degradation of multiple organic pollutants in water

Yurou Zhou[a], Guan Zhang[a,*] and Jing Zou[b]

[a] School of Civil and Environmental Engineering, Harbin Institute of Technology, Shenzhen (HITSZ), China
[b] General Education Division, School of Humanities & Social Science, The Chinese University of Hong Kong, Shenzhen (CUHKSZ), China
*Corresponding author. E-mail: zhangguan@hit.edu.cn

GZ, 0000-0001-7089-7536

ABSTRACT

In the present photoelectrochemical (PEC) system utilizing MoS_2 nanoflakes as a wide spectrum absorptive photoanode, simultaneous PEC degradation of different organic pollutants was achieved by employing *in situ* generated oxysulfur radicals, superoxide and hydroxyl radicals as strong oxidants. In order to better understand the cooperative PEC degradation of representative organic pollutants including rhodamine B dye (RhB), chlorpyrifos (CPF) and ciprofloxacin (CIP), the influences by bias potential, solution pH, radical scavenger, dissolved oxygen concentration and electrolyte concentration have been investigated. The selective PEC degradation efficiency follows the order of CPF > RhB > CIP in mixed substrates condition. In addition, the degradation rate for the single substrate degradation was about two times higher compared to that in mixed substrates degradation. The experimental results verified that reactive oxidation species (ROS) including oxysulfur radicals, superoxide and hydroxyl radicals can be efficiently produced on both anode and cathode under visible light irradiation, and they work together for simultaneous degradation of different pollutants, but the contribution of each ROS for pollutant degradation is substrate dependent. These results indicate that cooperative oxidation of multiple pollutants by miscellaneous oxygen-based radicals should be further considered as a promising advanced oxidation technique.

Key words: advanced oxidation, cooperative oxidation, multiple radicals, organics degradation, photoelectrocatalysis

HIGHLIGHTS

- MoS_2 nanoflakes on Ti sheet were prepared as a wide spectrum absorption anode.
- Sulfite ions served as electrolyte, precursor of oxysulfur radicals and hole scavenger.
- Multiple radical species can be produced by PEC activation of sulfite under visible light.
- Cooperative degradation of organic pollutants in wastewater was achieved.
- The contribution of each radical species for pollutant degradation was substrate-dependent.

GRAPHICAL ABSTRACT

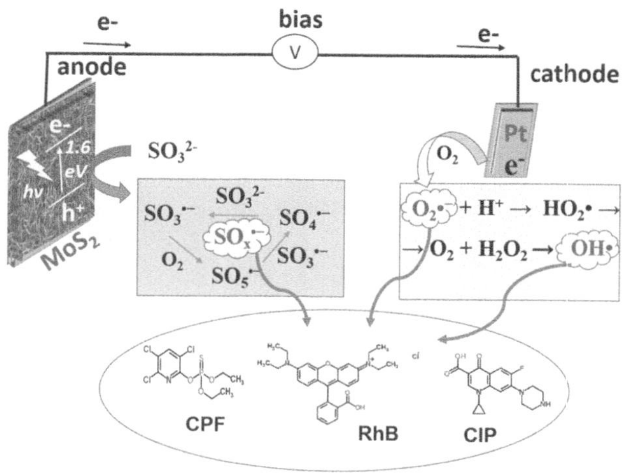

INTRODUCTION

Advanced oxidation processes (AOPs), as pretreatment approaches to increase biodegradability of wastewater or secondary treatment approaches after biological water treatment to further reduce the amount of organic pollutants, are increasingly employed for treatment of different types of organic wastewater. In addition, recalcitrant, high toxic, persistent and new emerging pollutants with low concentrations in wastewater, such as endocrine disrupting chemicals (EDCs), pharmaceuticals and personal care products (PPCPs) and persistent organic pollutants (POPs), are hardly removed by traditional physiochemical or biochemical approaches (Zhang *et al.* 2015; Huang *et al.* 2017). The AOPs utilize radical species (mainly hydroxyl radical, •OH) as strong oxidants for mineralization of different types of pollutants. Among the different types of AOPs (ozone oxidation, electro-catalytic oxidation, photo(electro)catalytic oxidation, Fenton oxidation, wet oxidation etc.), and photoelectrocatalytic (PEC) degradation of water pollutants has been broadly studied as one of the promising oxidation techniques (Pawar *et al.* 2015; Wu *et al.* 2015; Xia *et al.* 2017a, 2017b). However, whether in the photocatalytic or photoelectrocatalytic process, the search for abundant, stable and efficient photocatalytic materials is especially crucial from the perspective of practical application (Zhao *et al.* 2013; Ji *et al.* 2017; Liu *et al.* 2017a, 2017b).

On the other hand, even though hydroxyl radical has strong oxidation potential (2.8 V vs. SHE), it is not omnipotent since some kinds of water pollutants are resistant to its oxidation. For example, ammonia nitrogen is hardly completely removed by hydroxyl radicals (Deng & Ezyske 2011). Some research groups have reported conversion of ammonia nitrogen by photocatalysis, but most of the processes are less efficient (Wang *et al.* 2014; Bian *et al.* 2015; Dozzi *et al.* 2017; Liu *et al.* 2017a, 2017b). Our previous work also found that the molecular structures and properties of pollutants have a strong influence on the •OH mediated PEC degradation efficiency with the degradation rate order of chlorpyrifos > rhodamine B > 4-chlorophenol > ciprofloxacin (Zhou *et al.* 2019). Varanasi *et al.* (2018) also proposed that •OH selectively attack the aliphatic components, whereas Cl• induces transformation of olefinic species, and transformation of aromatic and olefinic moieties by $SO_4^{\bullet -}$ are the predominant pathway for removal of dissolved organic matter in UV–AOPs. Thus, the AOPs relying on hydroxyl radical alone are not sufficient for the handling of multiple recalcitrant pollutants. Besides the conventional •OH based AOPs, other radical species including superoxide radical ($O_2^{\bullet -}$), chlorine radical ($Cl^{\bullet -}$), oxysulfur radicals ($SO_3^{\bullet -}$, $SO_4^{\bullet -}$ and $SO_5^{\bullet -}$ radicals), persulfate radical ($S_2O_8^{\bullet -}$) and their combination mediated newly emerging AOPs have been paid more attention recently (He *et al.* 2014; Lutze *et al.* 2015; Remucal & Manley 2016; Jasper *et al.* 2017; Xia *et al.* 2017a, 2017b; Zhu *et al.* 2020; Ji *et al.* 2021). For instance, electrochemical and photoelectrochemical oxidation of ammonia by $Cl^{\bullet -}$ have been reported as efficient approaches for complete removal of ammonia (Yang *et al.* 2016; Shin *et al.* 2017). $SO_4^{\bullet -}$ radical in company with $Cl^{\bullet -}$ and •OH radicals-based water treatment has been studied (Lutze *et al.* 2015). In addition, an expanding range of sulfite, sulfate or persulfates activation methods have been established to generate oxysulfur radicals for degradation of organic pollutants (Farhat *et al.* 2015; Hu & Long 2016; Chen *et al.* 2017; Zhi *et al.* 2017).

To address the problems of AOPs as mentioned above, we attempt to explore a new PEC oxidation system based on multiple oxygen based radical species. Specifically, oxysulfur radicals are generated *in situ* at the photoanode, and O_2 reduction generated radicals like $O_2^{\cdot-}$ and •OH are produced at the cathode. These oxidative radicals can realize simultaneous degradation of different types of wastewater pollutants. Herein, MoS_2 nanoflakes grown on Ti sheet as a wide spectrum absorption photoanode was employed since it has a strong visible light absorption capability (band gap ~1.5–1.8 eV) and suitable valence band position (1.3–1.5 V vs SHE) for oxidation of sulfite to generate sulfite radical (Liu *et al.* 2016). Upon the analytical, microscopic and spectroscopic characterization of prepared MoS_2 electrodes, the degradation of three organic pollutants (chlorpyrifos as a representative of pesticide, rhodamine B as a representative of dye and ciprofloxacin as a representative of pharmaceutical) either in single substrate condition or mixed substrates condition was tested. These organic pollutants may coexist in some kinds of wastewater or sewage streams in different concentration levels, therefore the PEC degradation of three pollutants in mixed systems was studied. Furthermore, another goal of this work is to explore the molecular structure related degradation behavior, as the three pollutants have different molecular structures. Importantly, we have demonstrated that the *in situ* generated oxysulfur radicals, superoxide and hydroxyl radicals are key oxidants for the simultaneous degradation of different organic pollutants.

EXPERIMENTAL

Preparation and characterization of MoS_2 photoanode

The as-prepared MoS_2 nanoflakes on Ti film (MoS_2) and further annealed at 300 °C in reducing atmosphere (MoS_2(300 °C)) were obtained by a hydrothermal method, and detailed information can be found in our previous work (Zhou *et al.* 2019). The morphology, microstructure and crystal phases of MoS_2/Ti electrodes were characterized using a field emission scanning electron microscope (FESEM, SU8010), high resolution transmission electron microscope (HRTEM, Tecnai G2 F30), and X-ray diffraction (XRD, Bruker D8 Advance). Raman spectra of MoS_2 samples was measured with 532 nm excitation (LabRAMHR-800). Surface chemical composition and valence state of the samples were analysed by X-ray photoelectron spectroscopy (XPS, PHI 5000 VersaProbe II) with a monochromatic Al Kα ray source. The band-gap energies of different MoS_2/Ti films were measured by a UV-Vis spectrophotometer (JASCO V-7000).

Photoelectrochemical (PEC) performance evaluation

The PEC performance of different MoS_2 samples was tested by an electrochemical workstation (CHI660E, Shanghai Chenhua). Typically, the photocurrents of the MoS_2 photoanodes were recorded in 0.02 M Na_2SO_3 electrolyte and 0.5 V bias under visible light irradiation (≥420 nm) by a 300 W Xe arc lamp with a 420 nm filter (PLS-SXE300, light intensity 95.5 mW/cm^2) according to our previous optimization study (Zhou *et al.* 2019). The PEC degradation of rhodamine B (RhB) dye (2.0 μmol/ L) and chlorpyrifos (CPF) and ciprofloxacin (CIP) (10.0 μmol/L) was carried out with applied bias of 0.3–0.5 V in 0.01–0.05 M Na_2SO_3 electrolyte in the pH range of 3–11 utilizing an MoS_2(300 °C) electrode (electrode area = 1.0 cm^2). Continuous and slow air bubbling into the reactor aimed to provide higher amounts of oxygen for enhancing generation of radicals on the cathode and degradation efficiency. The concentration of RhB, CPF and CIP at different intervals was analysed by a UV–Visible spectrophotometer (Shimadzu UV-3600), gas chromatograph (GC, Fuli 9790) and high performance liquid chromatography (HPLC, Waters E2695). Detailed analytical information can be found in our previous work (Zhou *et al.* 2019). A series of radical trapping experiments were carried out by adding 5 vol.% methanol, tert-butyl alcohol (TBA) and 2,2,6,6-tetramethylpiperidine (TEMPO, 1 mM) as scavengers, respectively, in order to determine the role of each radical species.

RESULTS AND DISCUSSION

Physiochemical characterization of different MoS_2 electrodes

Figure 1(a) and Figure S1 (supplementary data) display the morphologies and microstructures of as-prepared MoS_2/Ti electrode (MoS_2) from hydrothermal synthesis and further annealed electrode at 300 °C (MoS_2(300 °C)). The vertical grown MoS_2 nanoflakes with a thickness of *ca.* 4–10 nm standing on Ti substrate were obtained by hydrothermal synthesis as shown in the SEM images. However, there are more cracks on the further annealed MoS_2 sample at 300 °C (Figure S1), probably because of the shrinkage and further growth of MoS_2, whereas the thickness of MoS_2 was not markedly varied. The length of vertically aligned MoS_2 layer on Ti substrate is about 1.2–2.0 μm for MoS_2 and MoS_2(300 °C) from the SEM image cross-section view,

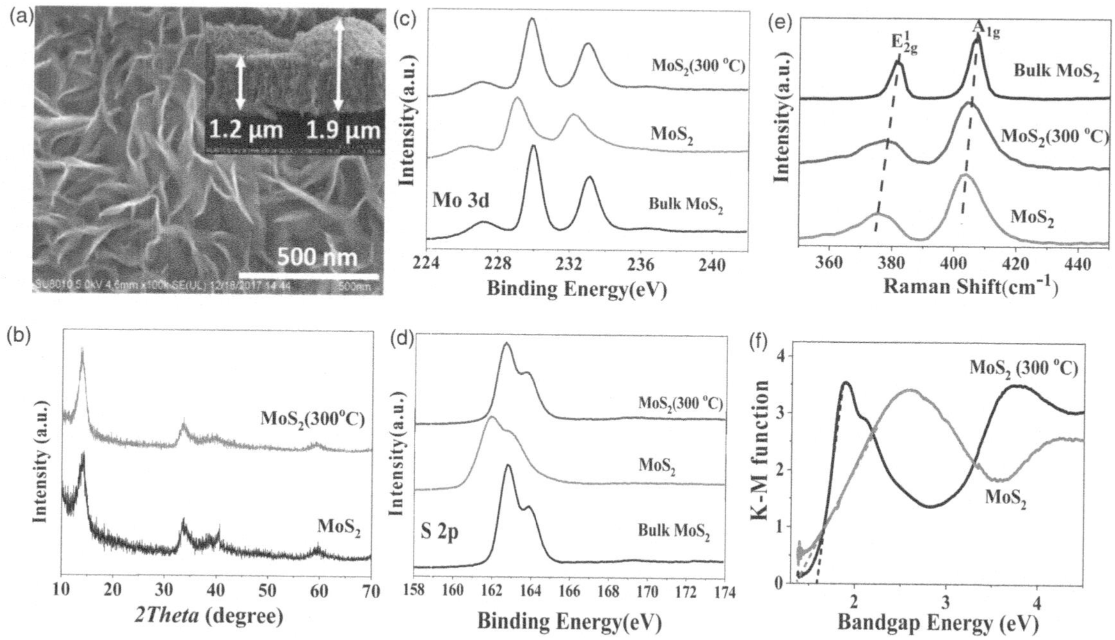

Figure 1 | (a) Top view and cross-section (inset) SEM images of MoS_2(300 °C) nanoflakes photoanode; (b) XRD patterns of different MoS_2 photoanodes; (c) Mo 3d XPS spectra of different MoS_2 photoanodse; (d) S 2p XPS spectra of different MoS_2 photoanodes; (e) Raman spectra of different MoS_2 photoanodes; (f) Diffused reflectance spectra of different MoS_2 photoanodes with K–M function plot.

as shown in the inset of Figure 1(a). In addition, the layered crystalline structure of MoS_2 can be identified from the HR–TEM images and SAED patterns (see Figure S2). The XRD patterns of different MoS_2 samples are shown in Figure 1(b). In reference to the standard MoS_2 diffraction peaks (JCPDS card: 73–1508), broad diffraction peaks at 14.4 and 34.1° with relatively low intensities are clearly exhibited for the MoS_2 sample. The MoS_2(300 °C) sample presents a higher diffraction peak of the (002) plane at 14.4° due to the enhanced crystallinity of MoS_2 nanoflakes after annealing at higher temperature.

Surface elemental XPS analysis of two prepared MoS_2 samples along with a commercial bulk MoS_2 sample (shown in Figure 1(c) and 1(d)) reflects some structural difference between the two prepared MoS_2 samples. The two peaks centred at ca. 230 and 233 eV (Mo $3d_{5/2}$ and Mo $3d_{3/2}$) for the bulk MoS_2 and MoS_2(300 °C) samples, and appeared at relatively higher binding energies compared to those of the as-prepared MoS_2 sample. Meanwhile, the S 2p doublet peaks for the bulk MoS_2 and MoS_2(300 °C) samples located at 163–164 eV, which is relatively higher than that of the as-prepared MoS_2. It was reported that Mo and S peaks of MoS_2 in 2H phase located at higher binding energy compared to 1 T phase MoS_2, indicating that structural transformation from 1 T phase to stable 2H phase occurred during the post annealing treatment. The representative Raman peaks of different MoS_2 samples are shown in Figure 1(e). The Raman peaks located at 383 and 409 cm^{-1} correspond to the in-plane ($E_{2\,g}$) and out-of-plane ($A_{1\,g}$) lattice vibrations. The Raman peaks of commercial bulk MoS_2 are narrow and sharp, whereas those for the prepared MoS_2 samples are broad, and there are slight Raman shifts for the prepared MoS_2 nanoflakes samples compared to the bulk MoS_2, probably due to edge effect and increased electronic transition energies in few-layered MoS_2. The band-gap energies of prepared MoS_2 film samples were evaluated by diffused reflection spectroscopy and the KM-plots of DRS spectra are shown in Figure 1(f). The band-gap energy of MoS_2 is associated with transitions from p-orbitals of the S atom to d-orbitals of the Mo atom in the visible light region (400–800 nm), and depends on physicochemical characters such as layer thickness and defects etc. The band-gap energies of the MoS_2 sample and MoS_2(300 °C) sample are about 1.35 and 1.60 eV, respectively. The increased band-gap energy of the MoS_2(300 °C) sample was probably due to the removal of intermediate defects, when the MoS_2 nanoflakes were annealed at higher temperature. The suitable band-gap energies of MoS_2 samples allow them to efficiently absorb solar light in a visible light region.

Photoelectrochemical measurements

Figure 2(a) exhibits the photocurrent response ($i-t$ curves) of different MoS_2 films under visible light irradiation at biased potentials of 0.5 V in 0.02 M Na_2SO_3 electrolyte. The photocurrent of MoS_2 was slightly reduced for the sample after

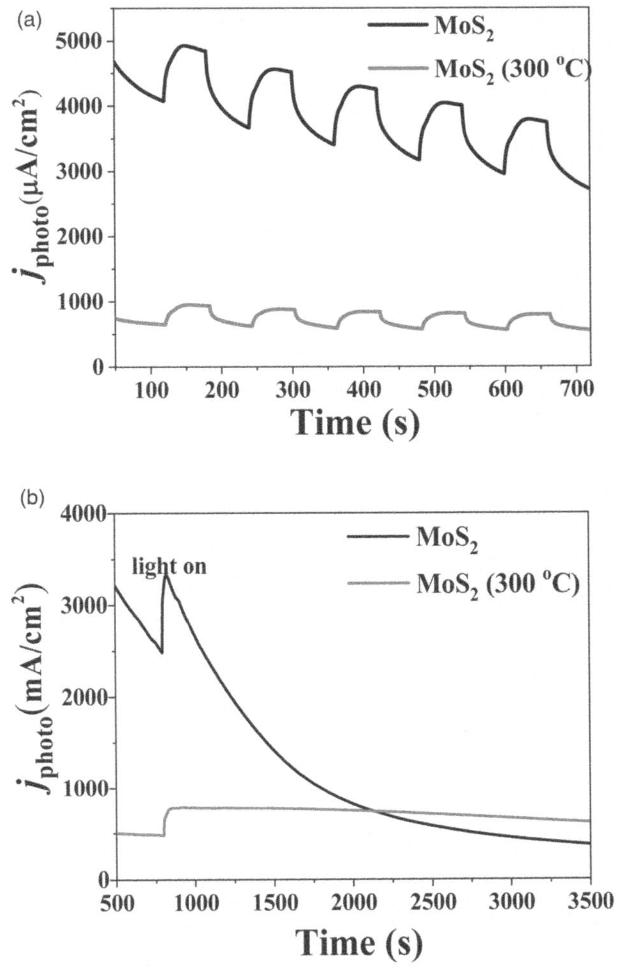

Figure 2 | (a) Chopped photocurrent–time ($i-t$) curves and (b) continuous photocurrent–time ($i-t$) curves of different MoS_2 photoanodes under 0.5 V bias (*vs.* Ag/AgCl) in 0.02 M Na_2SO_3 electrolyte.

annealing MoS_2 (300 °C). It was also found that the generated photocurrents are slowly increased and decreased during the cycled light 'on-off' processes. This slow photoresponse may be attributed to several aspects such as weak interfacial connection between Ti substrate and MoS_2 nanoflakes, intrinsic defects within MoS_2, and fast recombination of photogenerated charge carriers as evidenced from the physiochemical characterization. Furthermore, in the continuous test condition (Figure 2(b)), the as-prepared MoS_2 sample is extremely unstable, whereas the MoS_2(300 °C) sample is relatively more stable for the generation of photocurrents due to the post thermal treatment. Thus, the MoS_2(300 °C) sample was further selected for testing the PEC degradation of pollutants.

Three kinds of organic pollutants including rhodamine B dye (RhB), chlorpyrifos (CPF) and ciprofloxacin (CIP) (as shown in Table 1) were selected as representative pollutants in wastewater. Both a single substrate degradation experiment and mixed substrates degradation experiment were performed to investigate the selective degradation behavior of the MoS_2(300 °C) sample. The experimental parameters such as pH, bias potential, and electrolyte concentration were first studied for the degradation of mixed substrates, and the results are shown in Figure 3. The PEC degradation of three substrates was more favored in acidic conditions, as the PEC degradation rates were gradually increased when decreasing the solution pH from 11 to 3. Reducing the bias potential from 0.5 to 0.3 V does not markedly reduce the PEC degradation of the three substrates. However, increasing the electrolyte concentration from 0.01 to 0.05 M gradually improves the PEC degradation rate. From an energy and cost consumption point of view, the optimized PEC degradation condition was screened out with 0.3 V bias, pH 3 and 0.02 M Na_2SO_3. Under such optimized conditions, substrate-dependent degradation behavior can be clearly observed. The PEC degradation efficiency follows the order of CPF > RhB > CIP, as shown in Figure 4(a). The PEC

Table 1 | Molecular structures and degradation rate constants (*k*) at pH 3 during test conditions of single and mixed pollutants

Pollutant	Structure	Single pollutant		Mixed pollutants	
		Rate *k* (min^{-1})	**R^2**	**Rate *k* (min^{-1})**	**R^2**
Chlorpyrifos (CPF)	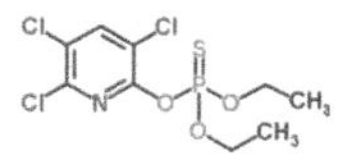	0.0369	0.98	0.0175	0.91
Rhodamine B (RhB)	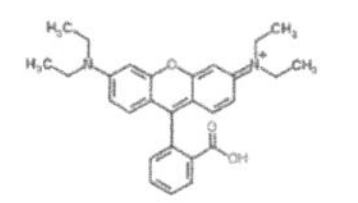	0.0259	0.97	0.0147	0.95
Ciprofloxacin (CIP)	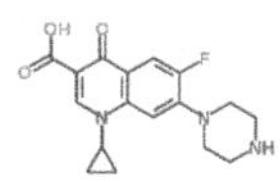	0.007	0.95	0.0037	0.88

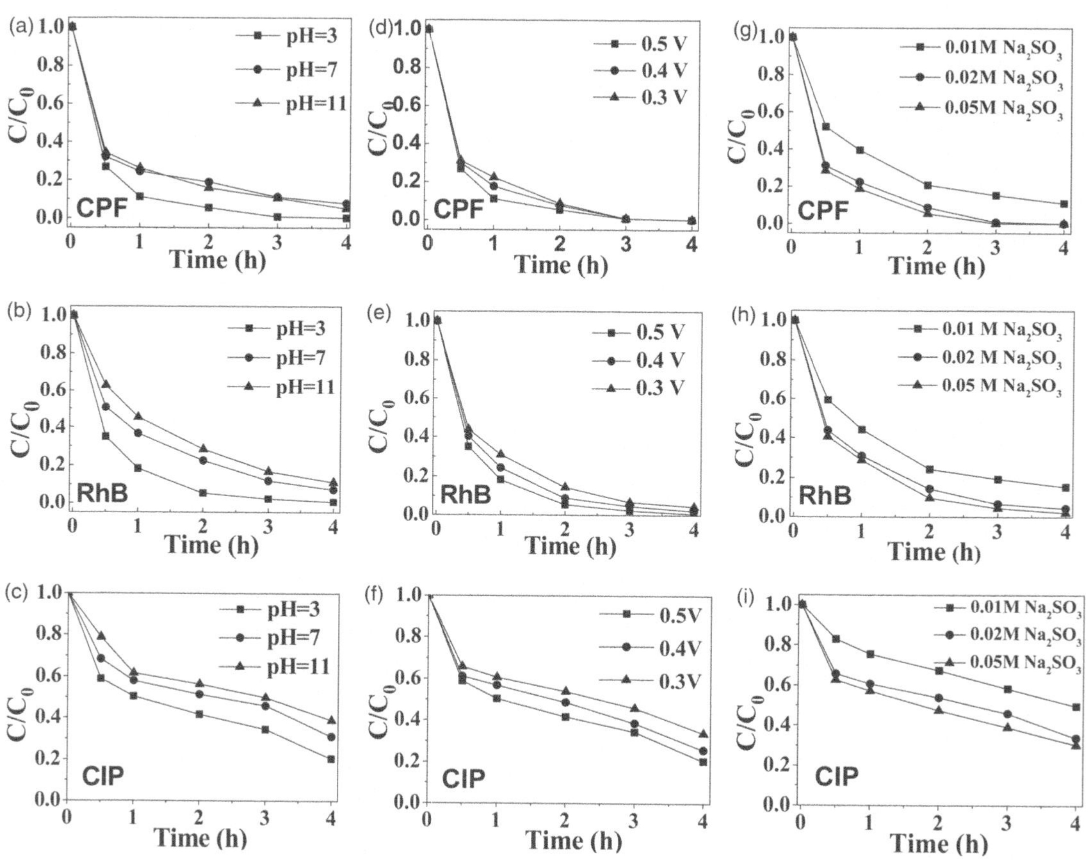

Figure 3 | (a)–(c) pH effect; (d)–(f) applied bias potential effect; and (g)–(i) electrolyte concentration effect on the photoelectrochemical degradation of rhodamine B dye (RhB), chlorpyrifos (CPF) and ciprofloxacin (CIP). Under varied experimental conditions: pH = 3–11, 0.3–0.5 V bias (*vs.* Ag/AgCl) and 0.01–0.05 M Na_2SO_3 electrolyte, [RhB] = 2 μmol/L, [CPF] = [CIP] = 10 μmol/L.

degradation rate constant of CPF was approximately five times higher than that of CIP (Figure 4(b) and Table 1), implying that CIP is a more chemically stable compound. In addition, a single substrate degradation experiment was performed and its degradation profile was compared with the mixed substrate condition (Figure 5). Generally, the degradation rate for the single substrate degradation was about two times higher compared to that of mixed substrates degradation (shown in Table 1). This is reasonable since there would be more radical species attacking pollutants in the single substrate condition. On the other hand, there would be more pollutants subjected to oxidation in the presence of mixed pollutants compared to

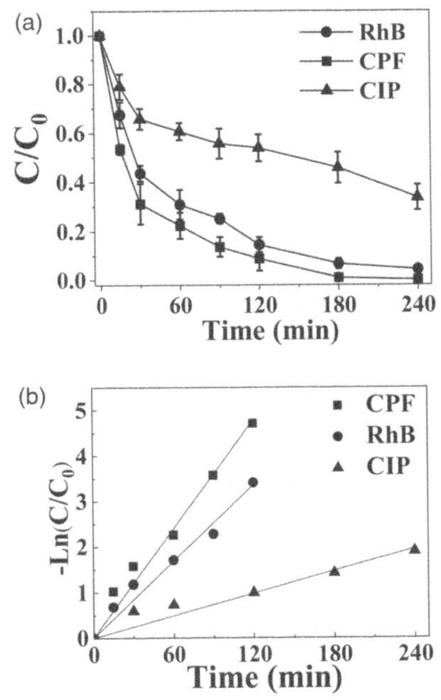

Figure 4 | (a) Profiles photoelectrochemical degradation of mixed three pollutants under optimized experimental conditions: pH = 3, 0.3 V bias (*vs.* Ag/AgCl) and 0.02 M Na_2SO_3 electrolyte; (b) comparison of first-order kinetic curves of photoelectrochemical degradation of mixed pollutants.

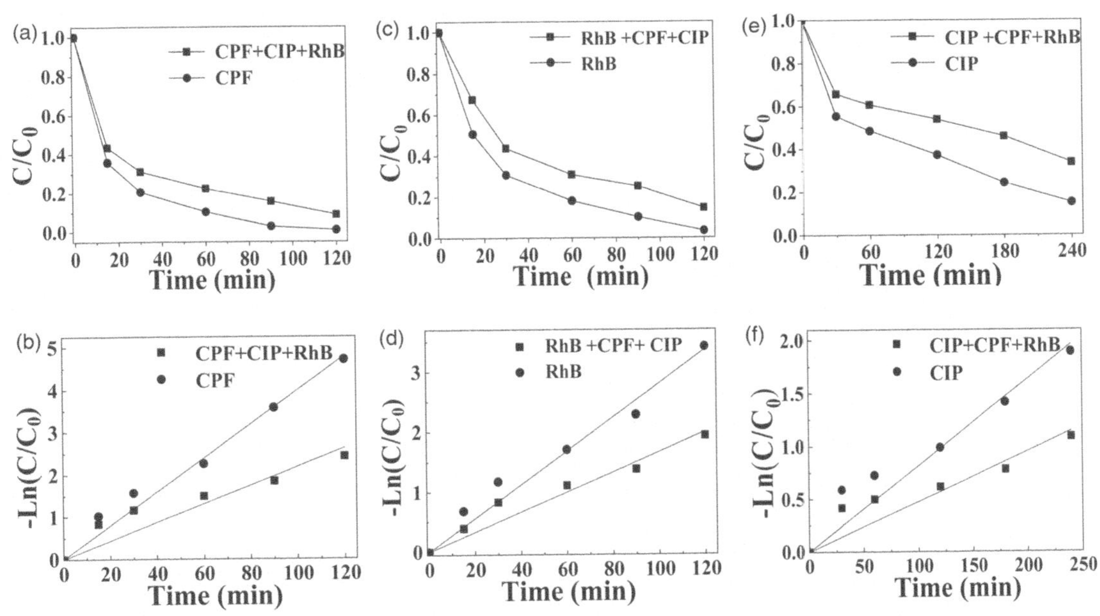

Figure 5 | (a) and (b) Chlorpyrifos (CPF) degradation profiles and first-order kinetic curves in test of single pollutant and mixed pollutants conditions; (c) and (d) Rhodamine B (RhB) degradation profiles and first-order kinetic curves in test of single pollutant and mixed pollutants conditions; (e) and (f) ciprofloxacin (CIP) degradation profiles and first-order kinetic curves in test of single pollutant and mixed pollutants conditions. Experimental conditions: pH = 3, 0.3 V bias (*vs.* Ag/AgCl) and 0.02 M Na_2SO_3 electrolyte.

that in the single pollutant conditions, because the recombination of generated oxidative radical species could be inhibited. In order to test the stability of the photoanode, multicycle experiments of RhB degradation were performed. As shown in Figure S3, after five cycle tests only slight PEC activity decay was observed, indicating the stability of the MoS_2(300 °C) sample. The stable characters of the MoS_2(300 °C) photoanode was attributed to the use of an Na_2SO_3 electrolyte that can consume photo-generated holes on MoS_2 to prevent self-corrosion.

Mechanism of PEC degradation of pollutants

The generation of reactive oxidative species (ROS) including •OH, $O_2^{\bullet-}$, and H_2O_2 on cathode and oxysulfur radicals ($SO_3^{\bullet-}$, $SO_4^{\bullet-}$ and $SO_5^{\bullet-}$) on the anode has been demonstrated in our previous work (Fan *et al.* 2019; Zhou *et al.* 2019). On the anode, the valence band potential (1.3–1.5 V *vs.* SHE) of MoS_2 is not sufficient for oxidizing water to generate •OH radicals, thus the self–oxidation of S^{2-} on the anode would be possible without hole scavengers. However, in the presence of sulfite electrolyte, the valence band holes generated on the nanostructured MoS_2 film under visible light irradiation are able to oxidize sulfite to sulfite radicals, because the redox potential of $SO_3^{\bullet-}/SO_3^{2-}$ couple is 1.2 V vs. SHE (Fan *et al.* 2019). Under such a setting, the self-oxidation of MoS_2 electrode was significantly inhibited. On the other hand, the SO_3^{2-} could also be electrochemically oxidized under biased conditions to produce $SO_3^{\bullet-}$. On the cathode, both the photochemically and electrochemically generated electrons would reduce O_2 to generate oxidative species like $O_2{\bullet}^{-}$, H_2O_2 and •OH. To inhibit the side reactions of H_2 production and enhance the generation of ROS, the reactor was continuously purged with air to feed higher amounts of oxygen. The oxygen effect on the degradation of pollutants can be seen in Figure 6. Without continuous air purging, only slight degradation of pollutants occurred for all the cases, probably because the dissolved oxygen was gradually depleted. This indicates that oxygen reduction-related ROS dominates the organics degradation.

To further prove the relative contributions of ROS taking part in the degradation of pollutants, quenching tests were carried out by adding different radical scavengers (methanol, tert-butyl alcohol (TBA), and TEMPO). TEMPO is a good scavenger of

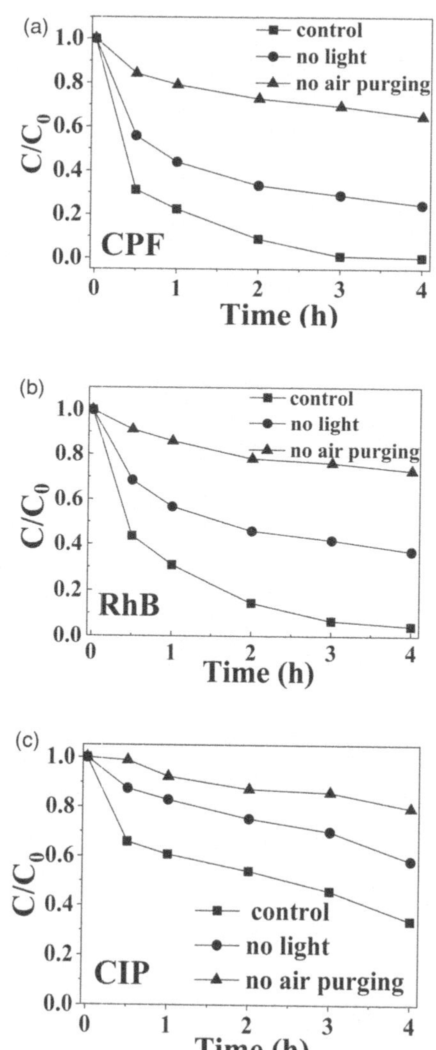

Figure 6 | Photoillumination effect and dissolved oxygen concentration effects on the photoelectrochemical degradation of three pollutants (a) CPF; (b) RhB; (c) CIP. Experimental conditions: pH = 3, 0.3 V bias (*vs.* Ag/AgCl) and 0.02 M Na_2SO_3 electrolyte.

radical species, methanol with an α–H can effectively quench oxysulfur radicals (e.g. $SO_4^{\bullet-}$, second-order rate constant 1.1×10^7 $Lmol^{-1}\,s^{-1}$), •OH (second-order rate constant 9.1×10^8 $Lmol^{-1}\,s^{-1}$) and holes, whereas TBA without an α–H mainly reacts with •OH (second-order rate constant 6×10^8 $Lmol^{-1}\,s^{-1}$) and is not effective for oxysulfur radicals (e.g. $SO_4^{\bullet-}$, 4×10^5 $Lmol^{-1}\,s^{-1}$). According to the different scavenging effects, the key ROS to each pollutant can be inferred. As shown in Figure 7(a), the scavenging effect from high to low in the CPF degradation according to the reaction rate is TEMPO (for •OH, $SO_x^{\bullet-}$, and $O_2^{\bullet-}$) > methanol (for •OH, $SO_x^{\bullet-}$) > TBA (for •OH). This result is consistent with the fact that oxygen reduction-related ROS played a critical role in the degradation of pollutants. The holes may be effectively scavenged by the sulfite electrolyte, thus the difference of methanol and TBA scavenging may be attributed to the $SO_x^{\bullet-}$ effect. For the RhB degradation test (Figure 7(b)), the TEMPO scavenging effect was excluded since its yellowish color has interference on the U-vis absorption of RhB solution. It can be seen that both methanol and TBA can effectively inhibit RhB degradation, indicating that •OH and $SO_x^{\bullet-}$ play a critical role in the RhB degradation. According to the scavenging effect, •OH could be the main ROS for RhB degradation. In the case of CIP degradation (Figure 7(c)), TEMPO and methanol present more significant scavenging effects compared to TBA, implying that $SO_x^{\bullet-}$ and $O_2^{\bullet-}$ radicals could be the key ROS for its degradation. Previously, we also found that •OH could not effectively degrade CIP, and Wang *et al.* (2018) reported that $O_2^{\bullet-}$ and photohole (h^+) from the C_3N_4 photocatalyst are the main oxidants for CIP degradation. The scavenging experiments prove that all the ROS, including •OH, $O_2^{\bullet-}$ and $SO_x^{\bullet-}$, participate in the organic degradation, as shown in Figure 8, but

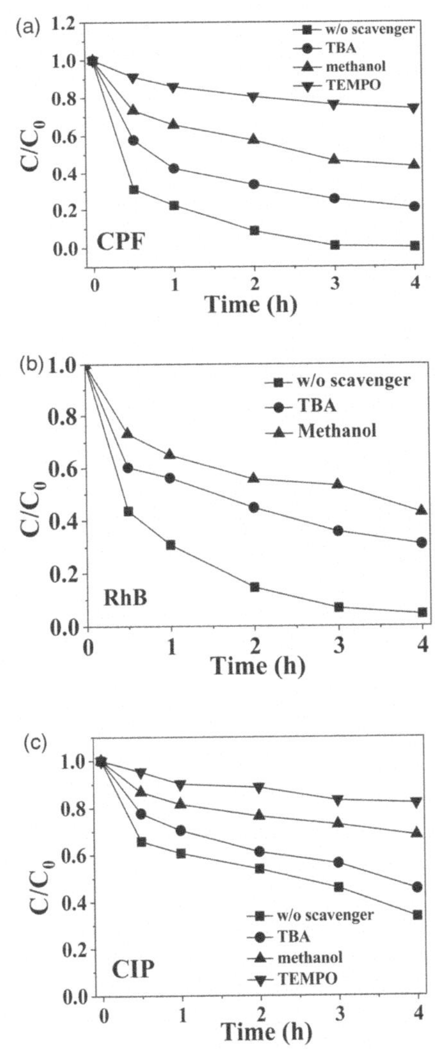

Figure 7 | Photoelectrochemical degradation of three pollutants: (a) chlorpyrifos (CPF); (b) rhodamine B dye (RhB); and (c) ciprofloxacin (CIP) in the presence of 5 vol.% tert-butyl alchol (TBA), methanol and TEMPO (1 mM) as ROS scavengers.

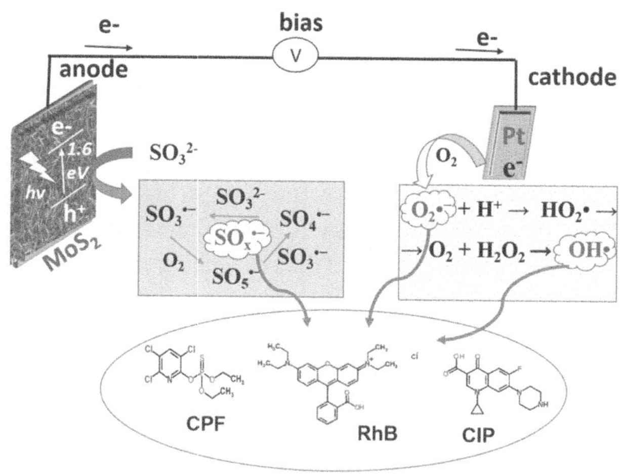

Figure 8 | Illustration of photoelectrochemical activation of sulfite for *in situ* generation of oxysulfur, superoxide and hydroxyl radicals towards simultaneous degradation of different pollutants.

the contribution of each ROS is substrate-dependent. However, oxygen reduction-related ROS generally contributes most to the pollutant degradation. The different oxidation characteristics of each ROS determine that they will preferentially attack different sites of pollutant molecules, and the degradation pathways and intermediates would be varied, which merits further study.

CONCLUSIONS

In summary, an efficient MoS_2–SO_3^{2-} based AOP employing oxysulfur radicals, hydroxyl and superoxide radicals together for the degradation of different structured pollutants has been clearly demonstrated. The coexisting radicals allowed more efficient PEC degradation of multiple organic pollutants with different structures. The PEC degradation rates were affected by many parameters such as applied bias, solution pH, dissolved oxygen concentration, and properties of the pollutants. Generally, acidic pH conditions and higher electrolyte concentrations were more favored for pollutants degradation. The PEC degradation efficiency was also substrate-dependent, with the general degradation rate order of chlorpyrifos > rhodamine B > ciprofloxacin in both single and mixed substrates conditions.

ACKNOWLEDGEMENTS

The authors are grateful for the financial support from NSFC (no. 51978197) and the Shenzhen Science and Technology Innovation Commission (Grant No. JCYJ20190813171403664), Basic research program of Guangdong Province (No. 2018A030313851).

DATA AVAILABILITY STATEMENT

All relevant data are included in the paper or its Supplementary Information.

REFERENCES

Bian, Z., Cao, F., Zhu, J. & Li, H. 2015 Plant uptake-assisted round-the-clock photocatalysis for complete purification of aquaculture wastewater using sunlight. *Environ. Sci. Technol.* **49**, 2418–2424.

Chen, L., Tang, M., Chen, C., Chen, M., Luo, K., Xu, J., Zhou, D. & Wu, F. 2017 Efficient bacterial inactivation by transition metal catalyzed auto-oxidation of sulfite. *Environ. Sci. Technol.* **51**, 12663–12671.

Deng, Y. & Ezyske, C. M. 2011 Sulfate radical-advanced oxidation process (SR-AOP) for simultaneous removal of refractory organic contaminants and ammonia in landfill leachate. *Water Res.* **45**, 6189–6194.

Dozzi, M. V., Brocato, S., Marra, G., Tozzola, G., Meda, L. & Selli, E. 2017 Aqueous ammonia abatement on Pt- and Ru-modified TiO_2: selectivity effects of the metal nanoparticles deposition method. *Catal. Today* **287**, 148–154.

Fan, X., Zhou, Y., Zhang, G., Liu, T. & Dong, W. 2019 In situ photoelectrochemical activation of sulfite by MoS_2 photoanode for enhanced removal of ammonium nitrogen from wastewater. *Appl. Catal. B Environ.* **244**, 396–406.

Farhat, A., Keller, J., Tait, S. & Radjenovic, J. 2015 Removal of persistent organic contaminants by electochemically activated sulfate. *Environ. Sci. Technol.* **49**, 14326–14333.

He, X., Cruz, A., O'Shea, K. E. & Dionysiou, D. D. 2014 Kinetics and mechanisms of cylindrospermopsin destruction by sulfate radical-based advanced oxidation processes. *Water Res.* **63**, 168–178.

Hu, P. & Long, M. 2016 Cobalt-catalyzed sulfate radical-based advanced oxidation: a review on heterogeneous catalysts and applications. *Appl. Catal. B Environ.* **181**, 103–117.

Huang, N., Wang, T., Wang, W. L., Wu, Q. Y., Li, A. & Hu, H. Y. 2017 UV/chlorine as an advanced oxidation process for the degradation of benzalkonium chloride: synergistic effect, transformation products and toxicity evaluation. *Water Res.* **114**, 246–253.

Jasper, J. T., Yang, Y. & Hoffmann, M. R. 2017 Toxic byproduct formation during electrochemical treatment of latrine wastewater. *Environ. Sci. Technol.* **51**, 7111–7119.

Ji, Y., Bai, J., Li, L., Xia, L., Chen, S. & Zhou, B. 2017 Improved photoelectrocatalytic activities of BiOCl with high stability for water oxidation and MO degradation by coupling RGO and modifying phosphate groups to prolong carrier lifetime. *Water Res.* **125**, 512–519.

Ji, J., Yan, Q., Yin, P., Mine, S., Matsuoka, M. & Xing, M. 2021 Defects on cos_{2-x}: tuning redox reactions for sustainable degradation of organic pollutants. *Angew. Chem. Int. Ed.* **60**, 2903–2908.

Liu, C., Kong, D., Hsu, P., Yuan, H., Lee, H., Liu, Y., Wang, H., Wang, S., Yan, K., Lin, D., Maraccini, P. A., Parker, K. M., Boehm, A. B. & Cui, Y. 2016 Rapid water disinfection using vertically aligned MoS_2 nanofilms and visible light. *Nature Nanotech.* **11**, 1098–1105.

Liu, S. Q., Zhu, X. L., Zhou, Y., Meng, Z. D., Chen, Z. G., Liu, C. B., Chen, F., Wu, Z. Y. & Qian, J. C. 2017a Smart photocatalytic removal of ammonia through molecular recognition of zinc ferrite reduced graphene oxide hybrid catalyst under visible light irradiation. *Catal. Sci. Technol.* **7**, 3210–3217.

Liu, S., Zhao, X., Zeng, H., Wang, Y. & Qiao, M. 2017b Enhancement of photoelectrocatalytic degradation of diclofenac with persulfate activated by Cu cathode. *Chem. Eng. J.* **320**, 168–177.

Lutze, H. V., Kerlin, N. & Schmidt, T. C. 2015 Sulfate radical-based water treatment in presence of chloride: formation of chlorate, inter-conversion of sulfate radicals into hydroxyl radicals and influence of bicarbonate. *Water Res.* **72**, 349–360.

Pawar, R. C., Pyo, Y., Ahn, S. H. & Lee, C. S. 2015 Photoelectrochemical properties and photodegradation of organic pollutants using hematite hybrids modified by gold nanoparticles and graphitic carbon nitride. *Appl. Catal. B Environ.* **176–177**, 654–666.

Remucal, C. K. & Manley, D. 2016 Emerging investigators series: the efficacy of chlorine photolysis as an advanced oxidation process for drinking water treatment. *Environ. Sci. Technol.* **2**, 565–579.

Shin, Y. U., Yoo, H. Y., Kim, S., Chuang, K. M., Park, Y. G., Huang, K. H., Hong, S. W., Park, H., Cho, K. & Lee, J. 2017 Sequential combination of Electro-Fenton and electrochemical chlorination processes for the treatment of anaerobically-digested food wastewater. *Environ. Sci. Technol.* **51**, 10700–10710.

Varanasi, L., Coscarelli, E., Khaksari, M., Mazzoleni, L. R. & Minakata, D. 2018 Transformations of dissolved organic matter induced by UV photolysis, hydroxyl radicals, chlorine radicals and sulfate radicals in aqueous-phase UV-Based advanced oxidation processes. *Water Res.* **135**, 22–30.

Wang, H., Su, Y., Zhao, H., Yu, H., Chen, S., Zhang, Y. & Quan, X. 2014 Photocatalytic oxidation of aqueous ammonia using atomic single layer graphitic-c_3n_4. *Environ. Sci. Technol.* **48**, 11984–11990.

Wang, F., Feng, Y., Chen, P., Wang, Y., Su, Y., Zhang, Q., Zeng, Y., Xie, Z., Liu, H., Liu, Y., Lv, W. & Liu, G. 2018 Photocatalytic degradation of fluoroquinolone antibiotics using ordered mesoporous g-C_3N_4 under simulated sunlight irradiation: kinetics, mechanism, and antibacterial activity elimination. *Appl. Catal. B Environ.* **227**, 114–122.

Wu, Z., Zhao, G., Zhang, Y., Liu, J., Zhang, Y. N. & Shi, H. 2015 A solar-driven photocatalytic fuel cell with dual photoelectrode for simultaneous wastewater treatment and hydrogen production. *J. Mater. Chem. A.* **3**, 3416–3424.

Xia, D., Yin, R., Sun, J., An, T., Li, G., Wang, W., Zhao, H. & Wong, P. K. 2017a Natural magnetic pyrrhotite as a high-efficient persulfate activator for micropollutants degradation: radicals identification and toxicity evaluation. *J. Hazard. Mater.* **340**, 435–444.

Xia, L. G., Bai, J., Li, J. H., Zeng, Q. Y., Li, L. S. & Zhou, B. X. 2017b High-performance $BiVO_4$ photoanodes cocatalyzed with an ultrathin α-Fe_2O_3 layer for photoelectrochemical application. *Appl. Catal. B Environ.* **204**, 127–133.

Yang, Y., Shin, J., Jasper, J. T. & Hoffmann, M. R. 2016 Multilayer heterojunction anodes for saline wastewater treatment: design strategies and reactive species generation mechanisms. *Environ. Sci. Technol.* **50**, 8780–8787.

Zhang, G., Ni, C., Liu, L., Zhao, G., Fina, F. & Irvine, J. T. S. 2015 Marcro-mesoporous resin as amorphous metal free visible light photocatalysts. *J. Mater. Chem. A* **3**, 15413–15419.

Zhao, X., Guo, L., Zhang, B., Liu, H. & Qu, J. 2013 Photoelectrocatalytic oxidation of Cu–EDTA at the TiO_2 electrode and simultaneous recovery of Cu by electrodeposition. *Environ. Sci. Technol.* **47**, 4480–4488.

Zhi, Y., Bai, J., Li, J., Luo, T., Qiao, L., Zeng, Q. & Zhou, B. 2017 Highly selective transformation of ammonia nitrogen to N_2 based on a novel solar-driven photoelectrocatalytic-chlorine radical reactions system. *Water Res.* **125**, 512–519.

Zhou, Y., Fan, X., Zhang, G. & Dong, W. 2019 Fabricating MoS_2 nanoflakes photoanode with unprecedented high photoelectrochemical performance and multi-pollutants degradation test for water treatment. *Chem. Eng. J.* **356**, 1003–1013.

Zhu, L., Ji, J., Liu, J., Mine, S., Matsuoka, M., Zhang, J. & Xing, M. 2020 Designing 3D-MoS_2 sponge as excellent cocatalysts in advanced oxidation processes for pollutant control. *Angew. Chem. Int. Ed.* **59**, 13968–13976.

First received 24 March 2021; accepted in revised form 16 June 2021. Available online 25 August 2021

© 2022 The Authors

Water Recycling and Low-Carbon Development doi: 10.2166/wrd.2021.046

Activation of peroxymonosulfate by $CuCo_2O_4$ nano-particles towards long-lasting removal of atrazine

Haoxiang Yin[a,b], Jun Li[c], Haoding Yan[b], Hanying Cai[b], Yanjian Wan[d], Gang Yao[b,e], Yong Guo[a,*] and Bo Lai[b]

[a] College of Chemical Engineering, Sichuan University, Chengdu 610065, China
[b] Sino-German Centre for Water and Health Research, Sichuan University, Chengdu 610065, China
[c] College of Water Conservancy and Hydropower Engineering, Sichuan University, Chengdu 610065, China
[d] Institute of Environmental Health, Wuhan Centers for Disease Prevention & Control, Wuhan 430022, China
[e] Institute of Environmental Engineering, RWTH Aachen University, Aachen 52072, Germany
*Corresponding author. E-mail: cdguoy@163.com

HY, 0000-0002-4344-1901; HC, 0000-0002-7234-2427

ABSTRACT

The effect of peroxymonosulfate (PMS) activated by nanocrystalline $CuCo_2O_4$ (NPS) on removal of atrazine (ATZ) was studied. First, the main experimental parameters were studied, including $CuCo_2O_4$ dose, PMS dose, initial pH value, and co-existing ion. The removal of ATZ (>99%) was attained under the optimal conditions (i.e., 150 mg/L $CuCo_2O_4$, 0.2 mM PMS, 5 mg/L ATZ, initial pH value of 6.8, and reaction time of 30 min). However, the removal of ATZ only reached 20.9% in the PMS alone system and there was no significant ATZ removal when adding $CuCo_2O_4$ alone into the solution, proving the good performance of the $CuCo_2O_4$/PMS system. Furthermore, reusability of $CuCo_2O_4$ was tested through five consecutive runs. To confirm which main active radicals were responsible in the system, two radical quenching experiments were carried out and electron paramagnetic resonance (EPR) was tested. In addition, the characterization of fresh and reacted $CuCo_2O_4$ NPs was tested by SEM, TEM, EDS, XRD, and XPS. Subsequently, based on the characterization of $CuCo_2O_4$ NPs and identification of radicals, $\equiv Cu^{2+}/\equiv Cu^{+}$ and $\equiv Co^{3+}/\equiv Co^{2+}$ were considered to be the main catalytic species, while the synergistic effect of Cu and Co played a crucial role. Finally, the degradation pathway of ATZ was proposed.

Key words: atrazine, catalytic mechanism, $CuCo_2O_4$, peroxymonosulfate

HIGHLIGHTS

- Nanoparticles $CuCo_2O_4$ was fabricated as peroxymonosulfate activator.
- ATZ could be effectively degraded by $CuCo_2O_4$/PMS system.
- Five degradation pathways of ATZ were proposed.
- $CuCo_2O_4$ has excellent reusability.

INTRODUCTION

For the last decade, the use of herbicides has become increasingly important in agriculture for weed control. As the most common used s-triazine herbicide, atrazine (ATZ) is widely applied for its high efficiency. Furthermore, ATZ has been applied for more than 40 years in over 80 countries (Hayes *et al.* 2002). However, because ATZ is heavily used in huge quantities, it has long been exposed to the environment. In addition, due to its high water mobility and persistence, traces of ATZ have been detected in surface water, even in groundwater (Khan *et al.* 2014; Li *et al.* 2018; Wang *et al.* 2020a). It will cause serious health problems, including impaired sexual development, deformities, and cancer (Hayes *et al.* 2002) when the concentration of ATZ exceeds 0.0001 mg/L (Pelizzetti *et al.* 1990). The application of ATZ has been prohibited by official order in many European countries (Ji *et al.* 2015), but its use still remains widespread in China (Chen *et al.* 2011). Therefore, it is necessary to continuously improve and develop efficient ATZ removal techniques.

In recent years, conventional advanced oxidation processes (AOPs), such as photocatalytic (Luo *et al.* 2015; Khaleel *et al.* 2019), Fenton (De Laat *et al.* 1999; Klamerth *et al.* 2009), Fenton-like (Zhang & Zhou 2019), and ozonation (Kuosa *et al.* 2009), have been developed to degrade organic pollutants in wastewater. All of the above methods can produce hydrogen radicals ($HO^{\cdot}$), whose redox potential is relatively high ($E^{\theta} = 2.8$ V). However, the deficiencies of the above treatment were non-negligible, including short life span of $HO^{\cdot}$, rigorous pH condition, and high cost. It has been found that $SO_4^{\cdot-}$ based on AOPs has a good performance on ATZ removal with a second-order rate constant of 10^6–10^9 $M^{-1}{\cdot}s^{-1}$ (Zhao *et al.* 2016a). As a radical of high redox potential ($E^{\theta} = 2.5$–3.1 V) (Zhang *et al.* 2019a; Nimai *et al.* 2020), $SO_4^{\cdot-}$ has a long-life span to contact with the targeted contaminant (Du *et al.* 2020). Moreover, because of its great selectivity (Li *et al.* 2018), $SO_4^{\cdot-}$ can react actively with compounds of unsaturated bonds and aromatic ring structures (Li *et al.* 2017). A common and efficient source of $SO_4^{\cdot-}$ is peroxymonosulfate (PMS), which has been successfully activated by many methods, such as transition metals (Anipsitakis & Dionysiou 2003; Zhou *et al.* 2016; Wang *et al.* 2020b), UV irradiation (Olmez-Hanci *et al.* 2011; Zhang *et al.* 2015), ultrasound (Lifka *et al.* 2003; Mahamuni & Adewuyi 2010; Zou *et al.* 2014), carbon-based catalysts (Chen *et al.* 2019), semiconductors (Jo *et al.* 2018), and so on.

Among those activation methods, transition metals have been regarded as one kind of low-cost and effective method, which is widely used in wastewater treatment to activate PMS. Transition metal catalysis can be divided into homogeneous and heterogeneous catalysis. The homogeneous catalysis often has high removal efficiency. For example, polychlorinated biphenyls (PCB) can be degraded by Fe^{2+}/PMS system with over 90% PCB removal within 24 h (Rastogi *et al.* 2009). However, the homogeneous catalysts for PMS activation have some shortcomings, including high metal ion dosage, production of metal oxide sludge, difficult to be separated and recovered from solution, and strict pH conditions (Li *et al.* 2017; Ding *et al.* 2019). Especially, for its high solubility and toxicity, Co^{2+} ion released into the solution after treatment will cause secondary pollution. Many countries have prohibited the application of Co^{2+} in wastewater treatment as a consequence.

In contrast to homogeneous catalysts, heterogeneous catalysts are easier to recycle from the solution. More importantly, heterogeneous catalysts usually produce little ion leaching, and, as a result, have been widely studied to activate PMS. $CuCo_2O_4$ spinel nanoparticles (NPs) have the characteristics of stable crystal structure, large specific surface area, low solubility, and are also highly efficient to activate PMS (Feng *et al.* 2015b). In fact, $CuCo_2O_4$ is widely used in supercapacitors, electrochemical sensors, and lithium batteries for its low cost, higher electronic conductivity, and excellent capacitance performance than either copper oxide or single Co (Liu *et al.* 2018; Xu *et al.* 2019). Notably, $CuCo_2O_4$ has excellent electrochemical properties, such as high electrical conductivity, which may contribute to electron transfer in AOPs to generate more radicals. Moreover, as a bimetallic oxide, $CuCo_2O_4$ has higher catalytic activity than monometallic oxide due to the synergistic catalytic effect of Cu and Co (Song *et al.* 2020). However, only a few studies have reported its application in the field of AOPs. Therefore, it is worth studying the removal of ATZ by $CuCo_2O_4$/PMS system.

In this work, the nanomaterial $CuCo_2O_4$, that can activate PMS efficiently, was studied. The main goals of the study are as follows: (i) to investigate the effects of main experimental parameters and inorganic anions on the removal of ATZ; (ii) to prove the performance of $CuCo_2O_4$/PMS system by two comparative experiments; (iii) to evaluate the stability and reusability of $CuCo_2O_4$ by five recycling experiments; (iv) to identify the free radicals which make a contribution to ATZ removal; (v) to deduce the catalytic mechanism from known results; (vi) to analyze the intermediate products of ATZ in the $CuCo_2O_4$/PMS system and infer the removal pathway of ATZ.

MATERIALS AND METHODS

Chemicals and reagents

Atrazine (≥97.0%), cobalt nitrate ($Co(NO_3)_2{\cdot}6H_2O$, ≥98.5%), copper nitrate ($Cu(NO_3)_3{\cdot}3H_2O$, ≥98.5%), copper oxide (CuO, ≥98.5%), cobalt sesquioxide (Co_2O_3, ≥98.5%), sodium bicarbonate ($NaHCO_3$, ≥98.5%), sodium nitrate ($NaNO_3$, ≥98.5%), sodium hydrogen phosphate ($NaHPO_4$, ≥98.5%), humic acid (HA, ≥90.0%), sodium chloride (NaCl, ≥98.5%), ethanol (EtOH), tert-butyl alcohol (TBA), and sodium hydroxide (NaOH, ≥98.0%) were purchased from Chengdu Kelong reagent company. Potassium peroxymonosulfate (PMS, $KHSO_5{\cdot}0.5KHSO_4{\cdot}0.5\ K_2SO_4$) was obtained from Sigma-Aldrich. The original solution of ATZ (5 mg/L) was obtained through dissolving ATZ with ultrapure water with stirring for 12 h. Ultrapure water (18.2 MΩ. cm) used in the experiments was produced by a UPH purification system (Ulupure, China).

Preparation of catalyst

In this study, $CuCo_2O_4$ nanoparticles were synthesized by sol-gel method (Zhao *et al.* 2016b). First, 0.025 mol $Co(NO_3)_2{\cdot}6H_2O$ and 0.05 mol $Cu(NO_3)_3{\cdot}3H_2O$ were dissolved in 100 mL deionized water and mechanically stirred at 300 rpm for 10 min. Then, 0.09 mol citric acid was added to the solution and stirred at 80 °C for 2 h. Next, after the mixture had been evaporated in a water bath of 90 °C for 1 h, the obtained gel complex was transferred to the oven and baked to a powder at the temperature of 90 °C. Lastly, the powder was transferred to a muffle furnace and heated to 550 °C at a heating rate of 5 °C/min, and eventually dried at the temperature of 550 °C for 5 h.

Experimental procedures

In general, all ATZ removal experiments were conducted with water bath heating, and a mechanical stirring speed of 300 rpm, reaction time of 30 min, and initial pH value of 6.8. At the beginning of the experiment, 150 mL of atrazine solution was transferred into a 400 mL beaker. Then, the beaker was placed in a water bath until the temperature of the water bath rose to 30 °C. Subsequently, the required amount of catalyst and oxidant were added at the same time. The samples were taken with syringes at predetermined intervals and filtered immediately with 0.45 μm PTFE syringe filter disks. Prior to sampling, as quenching agents of free radicals, 20 μL tert-butanol and ethanol were added to the sampling tube. Each data were the average of corresponding experiments repeated three times. Except for organic solvents, the medium of all other solutions is ultrapure water.

Analytical methods

FEI Tecnai G2 F20 S-TWIN (Hillsboro, Oregon, USA) scanning electron microscope (SEM) and transmission electron microscopy (TEM) were utilized to scan the surface morphology of the fresh $CuCo_2O_4$ NPs. The surface elementary compositions of $CuCo_2O_4$ NPs were obtained by energy dispersive spectrometer (EDS). The $CuCo_2O_4$ NPs were characterized by X-ray diffraction (XRD) EMPYREAN diffractometer (PANalytical B.V., The Netherlands). An X-ray photoelectron spectroscopy (XPS) (AXIS Ultra DLD, Kratos Co., UK) was used to characterize the chemical oxidation states on the surface of fresh and reacted $CuCo_2O_4$ NPs.

Atrazine concentration was detected by reversed-phase high performance liquid chromatography (Agilent USA) (Lai *et al.* 2018). A Bruker EMX-E spectrometer (Germany) was applied to monitor electron paramagnetic resonance (EPR) for detection of radicals. The intermediates of ATZ were analyzed by ultra performance liquid chromatography-quadrupole time of flight-mass spectrometer/mass spectrometer (UPLC-QTOF-MS/MS) system (Exion LC™ AC, X500R QTOF, SCIEX) (Hong *et al.* 2019). In the experiment, the removal of ATZ can be described by the pseudo-first-order kinetic pattern, which can be expressed as follows:

$$\ln (C/C_0) = -k_{\mathrm{obs}}t$$

where C is the concentration of ATZ of the sample taken at instant t, and C_0 is the initial atrazine concentration, k_{obs} is the kinetic rate constant, and t is the time of reaction.

RESULTS AND DISCUSSION

Characterization of $CuCo_2O_4$ NPs

SEM, TEM, EDS, and XRD analysis of $CuCo_2O_4$

The morphology and microstructure of $CuCo_2O_4$ NPs was characterized by scanning electron microscopy (SEM) and transmission electron microscopy (TEM). According to Figure 1(a) and 1(b), $CuCo_2O_4$ NPs showed an irregular layered structure with some nanoparticles on the surface. In addition, $CuCo_2O_4$ NPs were nanoparticles with diameters varying from 80 to 100 nm, measured by 'Nano measurer' software. Figure 1(c) shows the chemical atomic ratio of Co and Cu is 2.27:1. The percentage of O after usage increased for the increase of the absorbed H_2O on the $CuCo_2O_4$ NPs. Furthermore, $CuCo_2O_4$ NPs were characterized by XRD as shown in Figure 1(d). Figure 1(d) shows that the eight characteristic diffraction peaks at $2\theta = 19.0°, 31.4°, 36.8°, 39°, 44.5°, 55.5°, 59.6°, 65.2°$ can be indexed to (111), (220), (311), (222), (400), (422), (511), and (440) of spinel $CuCo_2O_4$ (Panayotov & Mehandjiev 1987), indicating the high purity of $CuCo_2O_4$. Moreover, the peak at 35.5° could be indexed to (−111) of CuO (JPCDS 48-1548), similar to the study by Wang *et al.* (2020c). The results were consistent with the removal of sulfamethazine by $CuCo_2O_4$/PMS system reported by Feng *et al.* (2015b). In addition, $CuCo_2O_4$ NPs exhibit a good performance in the structural stability since there was little difference and no new diffraction peak in the reacted particles, compared with the fresh $CuCo_2O_4$.

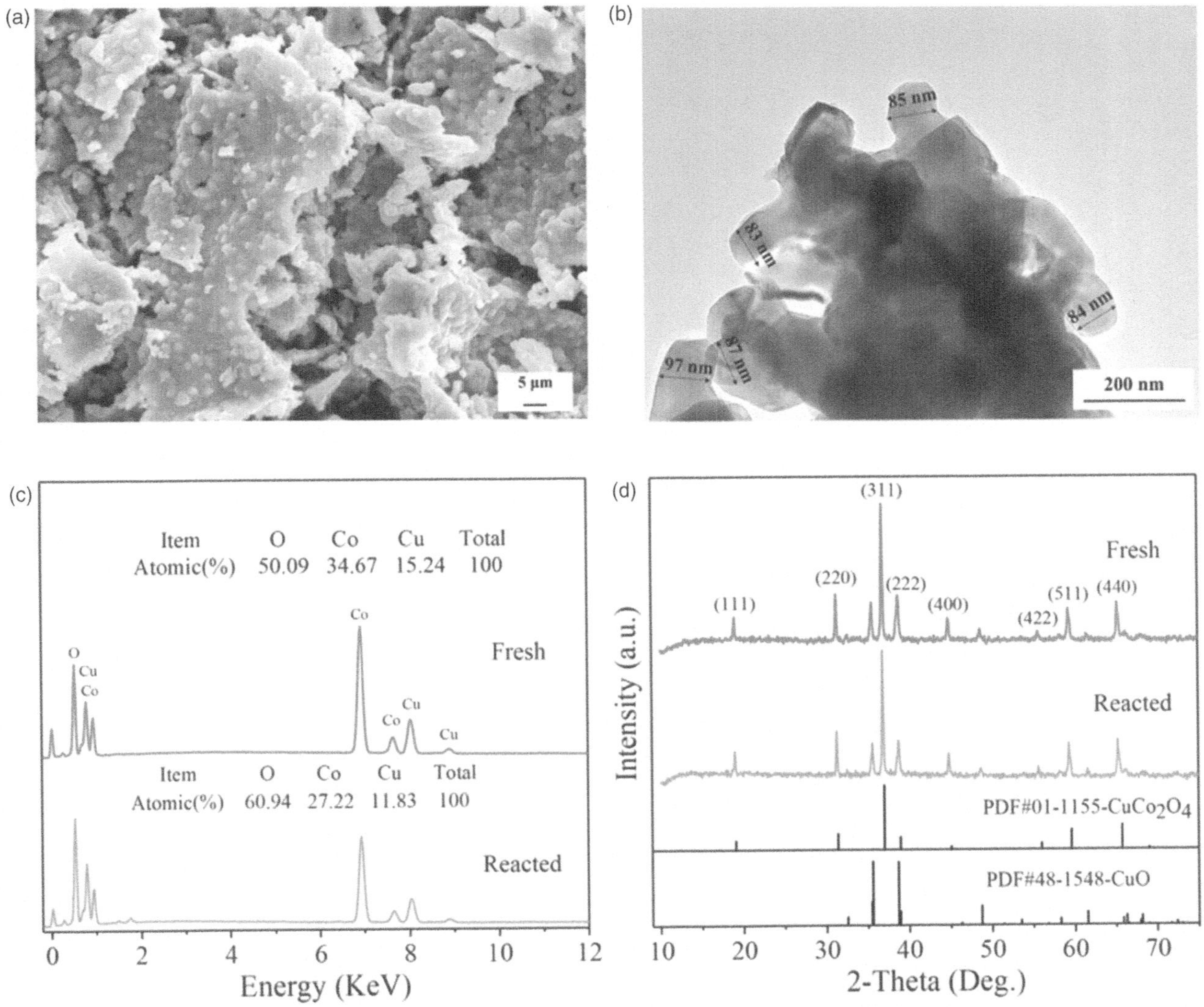

Figure 1 | (a) SEM, (b) TEM, (c) EDS, and (d) XRD of fresh and reacted $CuCo_2O_4$ NPs in $CuCo_2O_4$/PMS system for the removal of ATZ.

XPS analysis of $CuCo_2O_4$ NPs

To further understand the composition and chemical states of the elements on the surface of $CuCo_2O_4$, the XPS spectra of fresh and reacted $CuCo_2O_4$ NPs were acquired. Figure 2(a) shows a survey scan of $CuCo_2O_4$ NPs and illustrates the existence of Cu 2p, Co 2p, and O 1 s. Moreover, the binding energy was corrected by the carbon 1 s peak (284.8 eV) to obtain accurate spectra. Figure 2(b)–2(d) show the core level spectra of Co 2p, Cu 2p, and O 1 s.

The core-level spectra of Co 2p are shown in Figure 2(b) and four peaks at the binding energy of 780.45, 782.03, 795.54, and 797.30 eV were obviously observed. Corresponding to Co $2p_{3/2}$ and Co $2p_{1/2}$, respectively, both fitting peaks at 780.45 and 795.54 eV, which had a spin-energy separation of around 15 eV, indicated the existence of Co(II) (Wang *et al.* 2020c). In addition, the fitting peaks at 782.03 eV and 797.30 eV were assigned to Co(III) (Kibsgaard *et al.* 2012; Xu *et al.* 2018). In contrast to fresh $CuCo_2O_4$, the reacted $CuCo_2O_4$ NPs showed little difference in characteristic peak position and no new fitting peak was observed. As well, the peak area of Co(II) decreased from 45.44% to 42.25%, inferring the transformation between Co(II) and Co(III).

The XPS spectra of Cu 2p are displayed in Figure 2(c) to better understand the role of Cu in PMS activation process by $CuCo_2O_4$. Cu $2p_{3/2}$ was deconvoluted into two peaks with the binding energy of 934.08 eV and 937.17 eV, which were indexed into Cu(I) and Cu(II), respectively (Chen *et al.* 2020). Compared with the fresh $CuCo_2O_4$, the atom ratio Cu(I)/Cu(II) of reacted $CuCo_2O_4$ decreased from 7.29 to 7.00, which was supposed to be the conversion between Cu(I) and Cu(II).

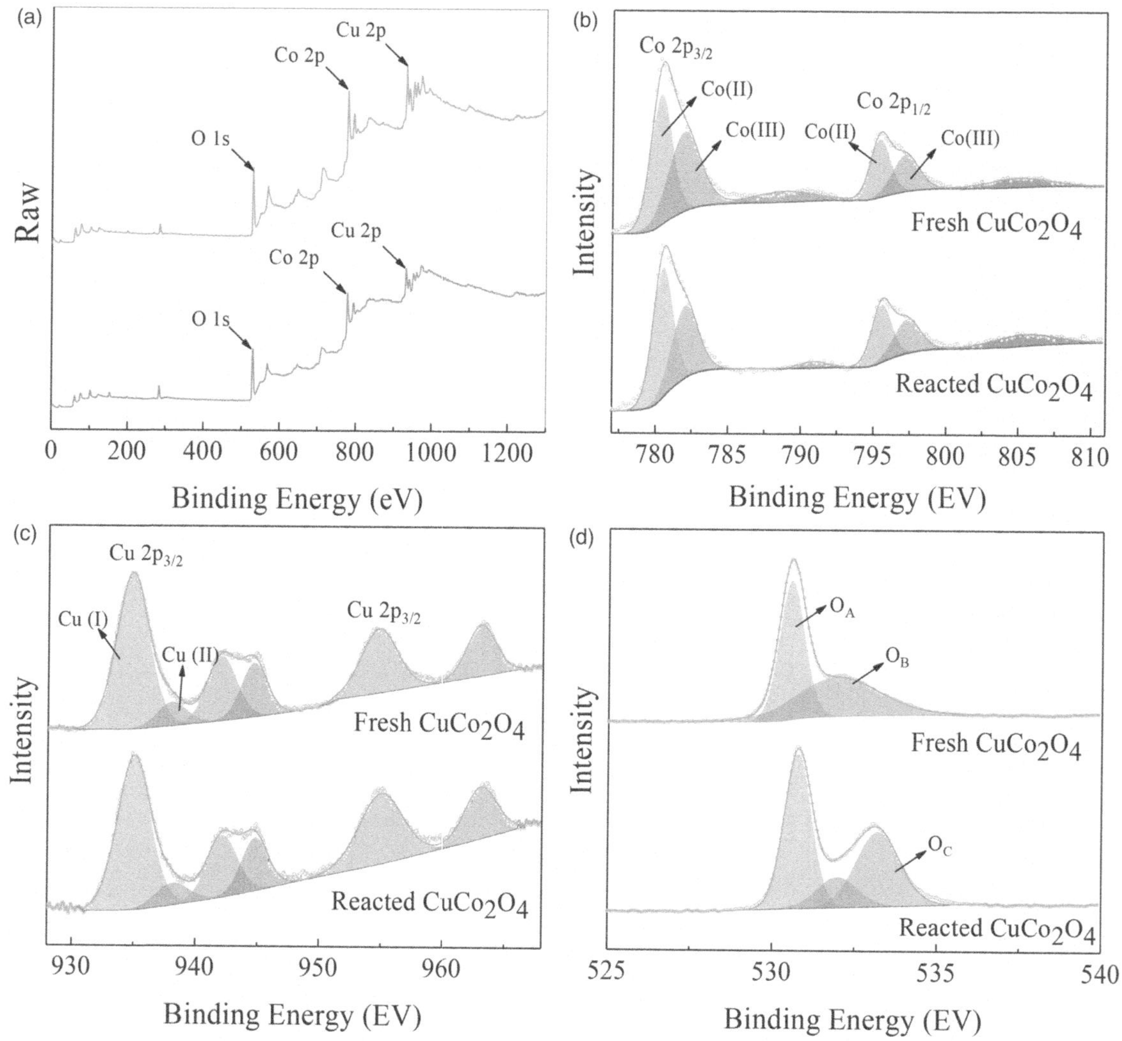

Figure 2 | XPS spectra of fresh and reacted $CuCo_2O_4$ NPs in $CuCo_2O_4$/PMS system: (a) full-range scan of the samples, (b) Co 2p core level, (c) Cu 2p core level, and (d) O 1 s core level.

Figure 2(d) depicts the core level spectra of O 1 s. The peak located at the binding energy of 530.61 eV was indexed into the lattice oxygen O_2^- (O_A) in $CuCo_2O_4$ NPs and the peak at around 532.01 eV belongs to the OH (O_B) adsorbed at the surface of $CuCo_2O_4$ (Xu *et al.* 2016). Compared with the fresh $CuCo_2O_4$, the area of the fitting peak at 532.01 eV decreased from 48.29% to 14.36% in the reacted $CuCo_2O_4$. Meanwhile, a new peak assigned to the O (O_C) of adsorbed H_2O appeared at the binding energy of 533.27 eV, suggesting that the hydroxyl groups on the surface of $CuCo_2O_4$ are converted into adsorbed water after the activation of PMS in the treatment process (Xu *et al.* 2016).

Analysis of BET and FT-IR

Analysis of BET and FT-IR is located in Text S1 in the Supplementary Material.

Effect of main experimental parameters

Effect of $CuCo_2O_4$ dose

The effect of $CuCo_2O_4$ NPs' dosage (0–400 mg/L) on the removal of ATZ was studied systematically. As seen from Figure 3(a) and 3(b), the ATZ removal was 7% when only PMS was added because of the weak oxidation capacity of PMS. However, with another addition of 20 mg/L $CuCo_2O_4$, the ATZ removal was enhanced significantly from 7% to 60% within 30 min. This manifests that PMS alone had a feeble effect on the removal of ATZ, and $CuCo_2O_4$ is superior in activating PMS. Moreover, the ATZ removal was highly dependent on the amount of catalyst when its addition was less than 150 mg/L. When its

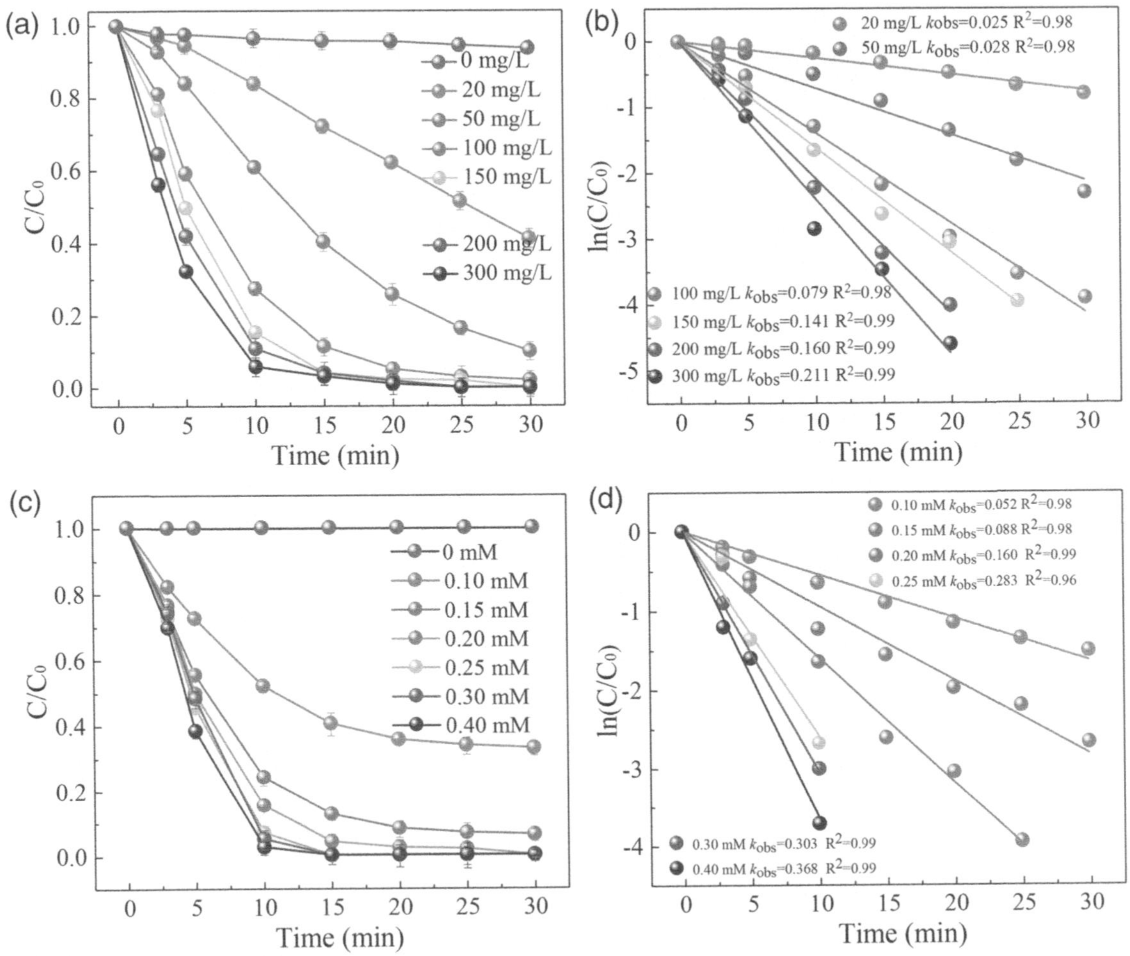

Figure 3 | Effect of ((a) and (b)) $CuCo_2O_4$ dosage and (c) and (d) PMS on the removal of ATZ. Experiment conditions: $[ATZ]_0 = 5$ mg/L, (a) and (b) $[PMS]_0 = 0.2$ mM, (c) and (d) $[CuCo_2O_4]_0 = 150$ mg/L, initial pH = 6.8, and reaction time = 30 min.

dosage increased to 150 mg/L, a removal efficiency of >99% was achieved. Meanwhile, the k_{obs} at 150 mg/L $CuCo_2O_4$ (0.160 min^{-1}) was nearly six times of that obtained at 20 mg/L $CuCo_2O_4$ (0.028 min^{-1}) in Figure 3(b). As more $CuCo_2O_4$ (>150 mg/L) was added, the removal efficiency remained at >99%. Its k_{obs} increased from 0.160 min^{-1} to 0.211 min^{-1}, 0.235 min^{-1}, 0.257 min^{-1} when the dosage of $CuCo_2O_4$ increased from 150 mg/L to 200 mg/L, 300 mg/L, and 400 mg/L. The reasons for this phenomenon were as follows: (1) the increased active site could promote the activation of PMS at the $CuCo_2O_4$ NPs dosage of 0–150 mg/L, leading to the enhancement of the removal of ATZ; (2) when the dose of $CuCo_2O_4$ exceeded 150 mg/L, $CuCo_2O_4$ was no longer the limiting factor for ATZ removal. However, the number of radicals produced per unit time increased with the dosage of $CuCo_2O_4$, increasing the k_{obs} value for ATZ removal. Therefore, the optimum dosage of $CuCo_2O_4$ was selected as 150 mg/L in the following experiments.

Effect of PMS dose

Figure 3(c) and 3(d) show the effect of PMS dosage (0 mM–0.40 mM) on ATZ removal. Compared with the ATZ removal by 150 mg/L $CuCo_2O_4$ alone, the removal of ATZ with the addition of 0.10 mM PMS and 150 mg/L $CuCo_2O_4$ significantly increased to 67% within 30 min of treatment. The phenomenon could be interpreted as follows. The $CuCo_2O_4$ almost had no adsorption on ATZ; (2) $CuCo_2O_4$ had a satisfactory performance in activating PMS in the $CuCo_2O_4$/PMS system. Furthermore, with the dose of PMS increasing from 0.10 mM to 0.20 mM, the ATZ removal was enhanced from 67% to >99%. When the addition of PMS exceeded 0.20 mM in the solution, the ATZ removal remained at >99%. For k_{obs} mentioned in Figure 3(d), when the dosage of PMS increased from 0.1 mM to 0.15, 0.20, 0.25, 0.30, and 0.40 mM, k_{obs} for ATZ removal increased from 0.052 min^{-1} to 0.088, 0.160, 0.283, 0.302, 0.368 min^{-1}, respectively. The following points could account for the results: (1) due to the direct oxidation by PMS, the increase of PMS dosage led to the enhancement of the removal of ATZ; (2) the active sites of $CuCo_2O_4$ had not been completely occupied by PMS, thus, with the increase of PMS, the contact between $CuCo_2O_4$ and PMS could be enhanced, producing more radicals to accelerate the ATZ removal. However, when more PMS (>0.2 mM) was added, the self-quenching phenomenon by excessive PMS reported by Shahzad *et al.* (2019) and Jawad *et al.* (2020) did not occur in the study. At the maximum additive amount of PMS, the excessive PMS in Shahzad's study exceeded 2 mM, which was ten times that in the study. This might be the reason why there was no self-quenching phenomenon in the $CuCo_2O_4$/PMS system. Therefore, the optimum dosage of PMS was selected as 0.2 mM in the following experiment.

Effect of initial pH

The effect of different initial pH (3, 5, 8, 9, and 11) on the removal of ATZ was studied (Figure 4(a) and 4(b)). Figure 4(a) shows that ATZ was barely removed at pH 11 and the highest ATZ removal (>99%) was reached at pH 5. Compared with the initial pH 5, the removal of ATZ at the initial pH 8 and 9 slightly declined to 94 and 85% and the corresponding k_{obs} for the removal of ATZ decreased from 0.205 to 0.100 and 0.061 min^{-1}, respectively. In addition, the ATZ removal significantly dropped to 9.8% when pH decreased to 3.0 and the k_{obs} declined from 0.205 to 0.003 min^{-1}.

Figure 4(c) shows the pH variation of the reaction solution at different initial pH. At the initial pH values of 5, 8, and 9, the pH of the solution dropped rapidly in the first 5 min due to the hydrolysis of PMS. For the next 25 min, the pH of the solution was maintained at 4.3, 5.5, and 6.4, respectively. When the initial pH value was 11, the pH of the solution decreased slightly

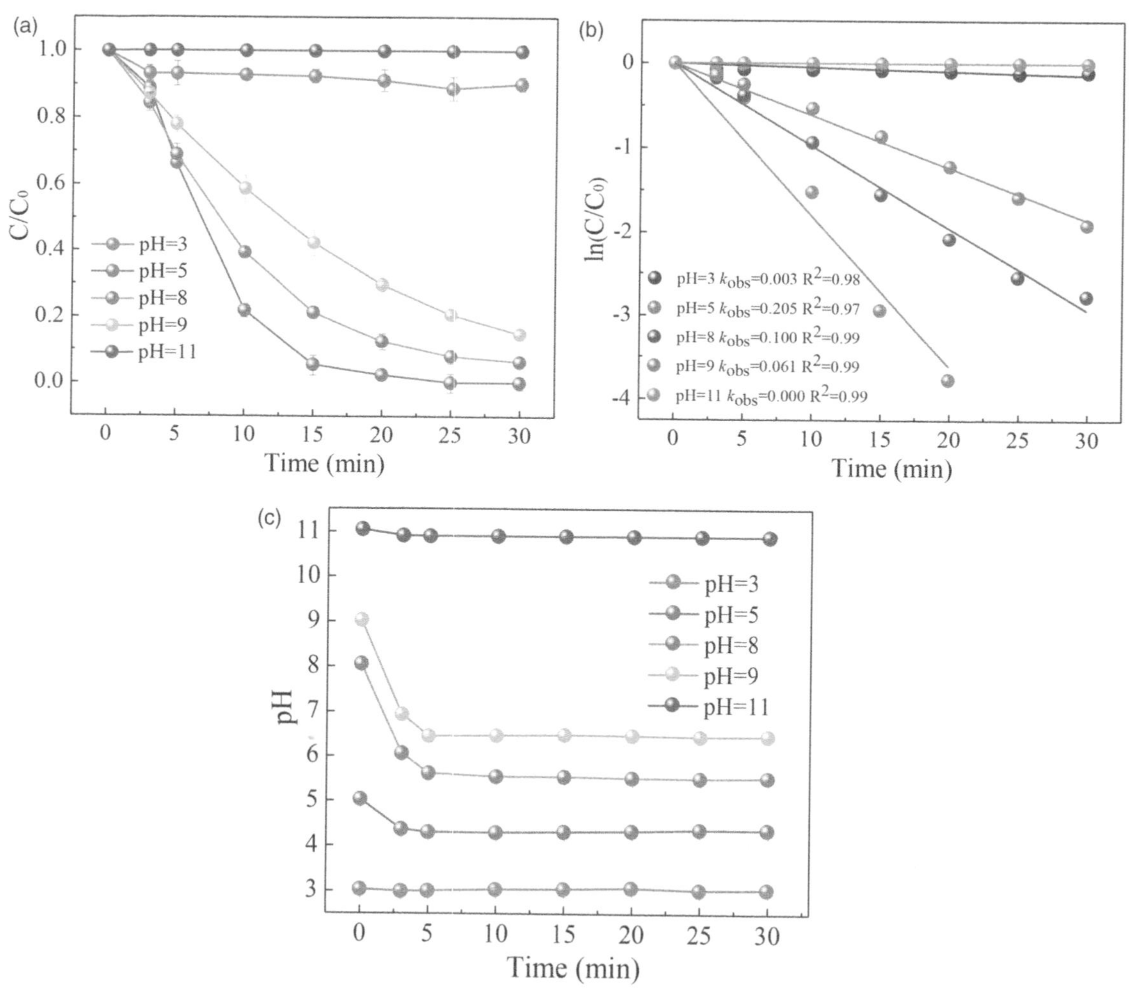

Figure 4 | Effect of (a and b) initial pH on the removal of ATZ, and (c) variation of solution pH during the reaction time. Experiment conditions: $[ATZ]_0 = 5$ mg/L, $[CuCo_2O_4]_0 = 150$ mg/L, $[PMS]_0 = 0.2$ mM, and reaction time = 30 min.

in the first 3 min and then barely changed. At the initial pH value of 3, the pH value remained almost constant during the process of ATZ removal.

According to the above results, weak acidic or neutral conditions are more favorable for the removal of atrazine, which was in line with the removal of diphenylamine in $PS/Fe^{3+}/Ag^{+}$ system (Li *et al.* 2009) and flumequine in PS/Fe_3O_4/MWCNTs/PHQ system (Feng *et al.* 2015a). The following factors can be used to interpret the phenomenon. (1) The zeta potential of $CuCo_2O_4$ at different pH (2–12) was detected. As illustrated in Supplementary Material, Figure S2, the zeta potential of $CuCo_2O_4$ showed a decreasing trend with the increase of pH. The results suggested that the higher the pH of the aqueous solution, the stronger the negative charge on the surface of $CuCo_2O_4$. PMS (HSO_5^- or SO_5^{2-}) was also negatively charged in the aqueous solution. Therefore, the electrostatic repulsion between PMS and $CuCo_2O_4$ was enhanced at a higher pH of the solution. Due to this factor, the ATZ removal underwent a drastic decline when pH increased from 5 to 8, 9, and 11. (2) $SO_4^{\bullet-}$ was identified as the dominant radical which was in favor of the removal of organic pollution under the acidic condition (Li *et al.* 2009). In alkaline condition, $SO_4^{\bullet-}$ would have an active reaction with OH^- (Equation (1)). This reaction could produce $HO^\bullet$ and SO_4^{2-}. The $HO^\bullet$ had a lower redox potential than $SO_4^{\bullet-}$ and the simultaneous production of SO_4^{2-} would inhibit the activity of $SO_4^{\bullet-}$ and $HO^\bullet$ (Deng *et al.* 2013). Therefore, ATZ removal was significantly reduced when the pH of the solution was alkaline. (3) An overly acidic environment (pH = 3) might also inhibit the ATZ removal since a large amount of H^+ could react with active species of $SO_4^{\bullet-}$ and $HO^\bullet$ to form HSO_4^- and H_2O (Equations (2) and (3)) (Zhang *et al.* 2019b). The pKa of ATZ was 1.6 (Chandra & Usha 2021). Therefore, more ATZ molecules present as neutral ones, which were relatively hard to be degraded by radicals than ATZ^- ions at pH 3. In addition, the leaching of Co and Cu was detected in Supplementary Material, Figure S3. When the pH decreased to 3, the obvious increase of the metal leaching was observed, suggesting that the synergistic effect of Cu and Co in $CuCo_2O_4$ NPs could be inhibited by the metal leaching of Cu and Co, which might be the reason for the decline in ATZ removal at lower pH (Chen *et al.* 2020). These factors might be the reason why ATZ removal dropped dramatically when pH value changes from 5.0 to 3.0.

$$SO_4^{\bullet-} + OH^- \rightarrow SO_4^{2-} + HO^\bullet \quad (1)$$

$$H^+ + SO_4^{\bullet-} + e^- \rightarrow HSO_4^- \quad (2)$$

$$H^+ + HO^\bullet + e^- \rightarrow H_2O \quad (3)$$

Effect of inorganic anions

The actual wastewater usually contains many kinds of anions and other compounds, including NO_3^-, HCO_3^-, HPO_4^{2-}, Cl^-, and humic acid. These inorganic anions and compounds might have a complex effect on the removal of ATZ. As a result, these five substances were added to ATZ solution, respectively, to investigate their effect on ATZ removal. The results are shown in Figure 5(a)–5(f).

The effect of HCO_3^- (0, 1, 10, and 25 mM) on the removal of ATZ was studied and the results are shown in Figure 5(a) and 5(b). With the addition of HCO_3^- increasing, the removal of ATZ decreased from >99% to 87.6%, 57.9%, 54.6%, and the corresponding k_{obs} value significantly declined from 0.160 to 0.070, 0.030, 0.027 min^{-1}, respectively. The results could be explained by the fact that HCO_3^- can react with $SO_4^{\bullet-}$ and $HO^\bullet$, producing $HCO_3^{\bullet-}$ (Equations (4) and (5)) (Ji *et al.* 2015; He *et al.* 2019) which had a lower redox potential. Supplementary Material, Figure S4 shows the pH change of $CuCo_2O_4$/PMS system during the reaction after the addition of HCO_3^-. As can be seen in Supplementary Material, Figure S4, when HCO_3^- was added to the $CuCo_2O_4$/PMS system, the solution became weakly alkaline, which was not conducive to the activation of PMS. This factor might be the reason for the inhibition in the ATZ removal.

$$HO^\bullet + HCO_3^- \rightarrow OH^- + HCO_3^{\bullet-} \quad (4)$$

$$SO_4^{\bullet-} + HCO_3^- \rightarrow HCO_3^{\bullet-} + SO_4^{2-} \quad (5)$$

The influence of the NO_3^- (0, 1, 10, and 25 mM) inorganic anion on the removal of ATZ was investigated in Figure 5(c) and 5(d). The ATZ removal remained at >99% with little change of the k_{obs} when the concentration of NO_3^- increased from 0 to 25 mM. This result reveals that concentration of NO_3^- less than 25 mM had a slight influence on the ATZ removal.

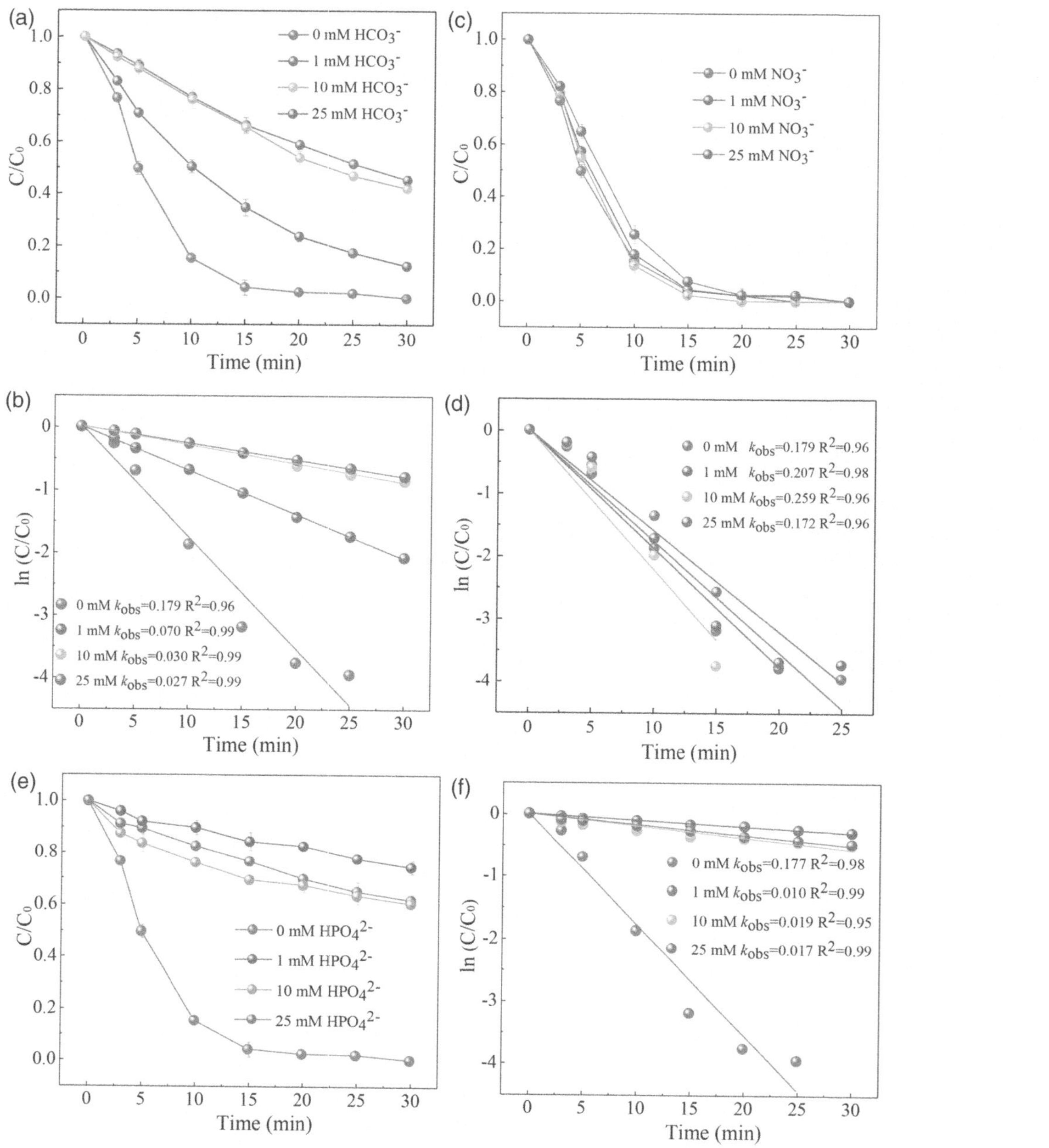

Figure 5 | Influences of (a and b) HCO_3^-, (c and d) NO_3^-, (e and f) HPO_4^{2-}, (g and h) Cl^-, and (i and j) HA on the removal of ATZ. Experiment conditions: $[ATZ]_0 = 5$ mg/L, $[CuCo_2O_4]_0 = 150$ mg/L, $[PMS]_0 = 0.2$ mM, initial pH = 6.8, and reaction time = 30 min. (*Continued.*)

Furthermore, NO_3^- at the concentration of 25 mM showed little inhibition of the removal of ATZ. The results could be interpreted as ATZ removal is not affected by a small amount of NO_3^- in the $CuCo_2O_4$/PMS system, which is consistent with the degradation of ATZ by the $CoFe_2O_4$/PMS system (Li *et al.* 2018).

Figure 5(e) and 5(f) show the influence of HPO_4^{2-} (0, 1, 10, and 25 mM) on the removal of ATZ. Compared to no HPO_4^{2-} addition, the ATZ removal decreased from >99% to 25.6% with adding 1 mM HPO_4^{2-} to the reaction solution in the $CuCo_2O_4$/PMS system. Meanwhile, the k_{obs} for ATZ removal showed a distinct reduction (from 0.160 to 0.009 min^{-1}) when the addition concentration of HPO_4^{2-} changed from 0 to 1 mM. Nevertheless, the ATZ removal did not show a further decrease when the addition concentration of HPO_4^{2-} increased from 1 to 10 and 25 mM. On the contrary, the ATZ removal increased from 25.6% to 38.2% and 39.5% and the k_{obs} also increased slightly from 0.09 min^{-1} to 0.015 min^{-1} and

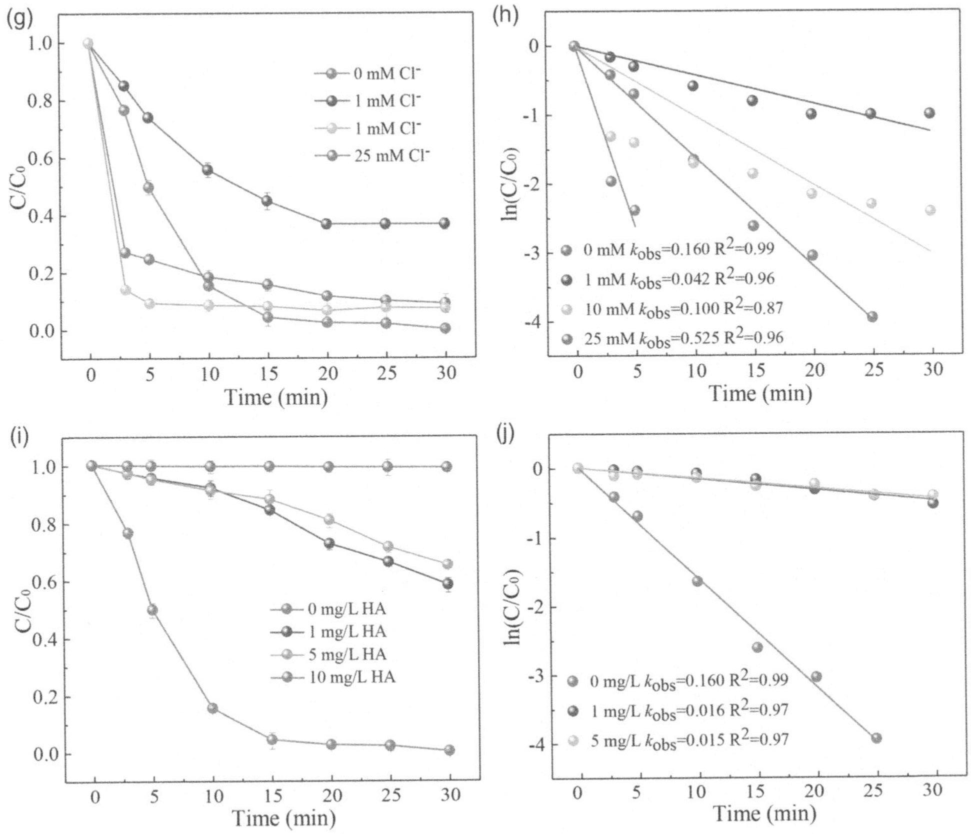

Figure 5 | Continued.

0.016 min^{-1}, respectively. In brief, with increasing the addition concentration of HPO_4^{2-}, the inhibition of ATZ removal increased initially, followed by a decrease. The inhibition of ATZ removal could be explained as $SO_4^{\bullet-}$ and $HO^{\bullet}$ would react with HPO_4^{2-} and produce some species with lower redox potential according to Equations (6) and (7) (Buxton *et al.* 1988; Neta *et al.* 1988). The reason for the reduction of the inhibition was as follows: when HPO_4^{2-} was excessive, some of the HPO_4^{2-} could scavenge $SO_4^{\bullet-}$ and $HO^{\bullet}$, and the rest would activate PMS to produce $SO_4^{\bullet-}$ and $HO^{\bullet}$. The activation of HPO_4^{2-} for PMS was demonstrated in a previous study by Lou *et al.* (2014). Therefore, the inhibition of ATZ removal was still in existence but reduced when the concentration of HPO_4^{2-} was higher than 1 mM.

$$HO^{\bullet} + HPO_4^{2-} \rightarrow HPO_4^{\bullet-} + OH^- \tag{6}$$

$$SO_4^{\bullet-} + HPO_4^{2-} \rightarrow HPO_4^{\bullet-} + SO_4^{2-} \tag{7}$$

The effect of different concentrations of Cl^- (0, 1, 10, and 25 mM) on ATZ removal was investigated. As exhibited in Figure 5(g) and 5(h), low concentration of Cl^- significantly inhibited the removal of ATZ. On the contrary, an increase in the concentration of Cl^- decreased the degree of inhibition. When the concentration of Cl^- increased to 25 mM, the k_{obs} for ATZ removal increased to 0.525 min^{-1}, suggesting that high concentration of Cl^- promoted the removal of ATZ to a certain extent. The results can be explained as follows: (1) at low concentration, Cl^- would compete with ATZ for $SO_4^{\bullet-}$ and $HO^{\bullet}$ to generate less reactive chlorine species such as $Cl^{\bullet}$ (2.4 V) and $Cl_2^{\bullet-}$ (2.1 V) (Equations (8)–(10)), causing the reduction of the removal of ATZ; (2) when the concentration of Cl^- increased to 25 mM, Cl^- would react with PMS to form HClO (Equation (11)). Although the redox potential of HClO (1.63 V) was low, HClO had a much longer lifetime as

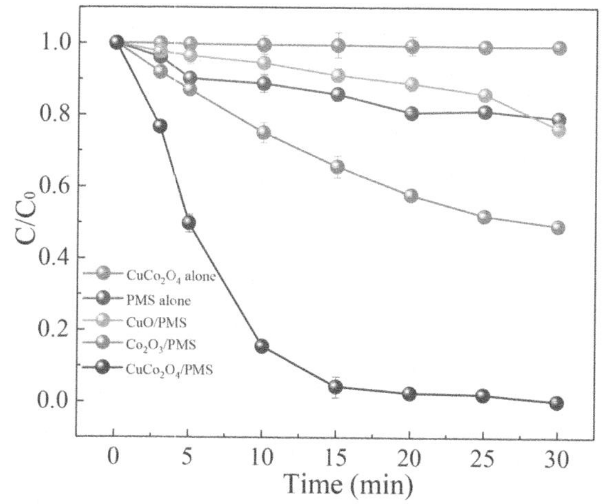

Figure 6 | ATZ removal in the different systems. Experiment conditions: $[ATZ]_0 = 5$ mg/L, $[CuCo_2O_4]_0 = 150$ mg/L, $[CuO]_0 = 49$ mg/L, $[Co_2O_3]_0 = 101$ mg/L, $[PMS]_0 = 0.2$ mM, initial pH = 6.8, and reaction time = 30 min.

compared to radicals, and can become the main oxidant in high Cl^- conditions (Lutze *et al.* 2015; Lee *et al.* 2020).

$$HO^{\bullet} + Cl^- \rightarrow Cl^{\bullet} + OH^- \tag{8}$$

$$SO_4^{\bullet -} + Cl^- \rightarrow Cl^{\bullet} + SO_4^{2-} \tag{9}$$

$$Cl^{\bullet} + Cl^- \rightarrow Cl_2^{\bullet -} \tag{10}$$

$$HSO_5^- + Cl^- \rightarrow SO_4^{2-} + HClO \tag{11}$$

As the model compounds of natural organic matter, different concentrations of humic acid (0, 1, 5, and 10 mg/L) were introduced to study its effect on ATZ removal (Figure 5(i) and 5(j)). An apparent inhibition in the removal of ATZ was observed. With the increase of HA, the inhibition for the removal of ATZ also increased. When the addition of HA increased from 0 mM to 1, 5, and 10 mg/L, the removal of ATZ decreased from >99% to 41.9%, 34.5, and 0%. Similar results were also found in Sharma *et al.* (2015) study. HA has often been applied as quencher for free radicals (Sharma *et al.* 2015). Therefore, HA may compete with ATZ for $SO_4^{\cdot -}$ and $HO^{\cdot}$, resulting in the decrease of the removal of ATZ.

Control experiments

To prove the activation performance of the $CuCo_2O_4$/PMS system, PMS alone, $CuCo_2O_4$ alone, CuO/PMS, and Co_2O_3/PMS systems were set up. Figure 6 shows that the removal of ATZ can achieve 20.9% when only PMS was present in aqueous solution. The results inferred that ATZ could be removed by PMS for direct oxidation. Meanwhile, the 150 mg/L $CuCo_2O_4$ alone system had little adsorption to ATZ. Nevertheless, the removal of ATZ was signally improved in the $CuCo_2O_4$/PMS system. The above results indicated the $CuCo_2O_4$/PMS system had a significant effect on the removal of ATZ because PMS activated by $CuCo_2O_4$ produced large amounts of radicals (i.e., $SO_4^{\cdot -}$ and HO·). Moreover, the performance of CuO and Co_2O_3 was evaluated under the same conditions. As can be seen in Figure 6, the ATZ removal by CuO/PMS could achieve 23.7% and that by Co_2O_3/PMS could reach 50.9%, both of which were far lower than with $CuCo_2O_4$/PMS. The results suggested that $CuCo_2O_4$ NPs had a better performance in catalytic activity than CuO or Co_2O_3 alone.

As shown in Supplementary Material, Figure S5, the pH value in the three systems was recorded. The pH value showed slight fluctuations when only $CuCo_2O_4$ was added to the aqueous solution. However, the pH decreased to 4.00 and 4.30 in PMS alone and $CuCo_2O_4$/PMS system, respectively, which was contributed to by the production of amount of H^+ according to Equations (12) and (13):

$$HSO_5^- + H_2O \rightarrow H^+ + SO_5^{2-} \tag{12}$$

$$HSO_5^- + H_2O \rightarrow H_2O_2 + HSO_4^- \tag{13}$$

Actual application

To study the ATZ removal ability of the $CuCo_2O_4$/PMS system in actual wastewater, four ATZ solutions were prepared with ultrapure water, tap water, river water, and lake water, respectively. The samples of tap water and lake water were taken from Sichuan University, and river water was taken from the Fuhe river in Chengdu. Supplementary Material, Table S1 shows the characteristics of various water matrices. The river and lake water were suction filtered and then used as the solvent for atrazine. The removal of ATZ in these four solutions by $CuCo_2O_4$/PMS system is illustrated in Supplementary Material, Figure S6. The removal of ATZ reduced in order of deionized water, tap water, river water, and lake water. However, the biodegradability of all solutions after the reaction was improved. As shown in Supplementary Material, Figure S7, the biodegradability of lake water increased slightly from 0.820 to 0.848 (expressed as BOD/COD). This might be because the actual water contained a large number of pollutants that could compete with ATZ for radicals.

Reusability of $CuCo_2O_4$ NPs

Supplementary Material, Table S2 shows the leaching concentration of cobalt and copper iron during reaction in the $CuCo_2O_4$/PMS system. The leaching concentration of the cobalt and copper ion increased with the increase of reaction time. The maximum leaching of Co and Cu was 0.035 mg/L and 0.062 mg/L at 30 min, respectively, which was much lower than the limit concentration (1 mg/L) required by *Emission standard of pollutants for copper, nickel, cobalt industry of China* (GB 25467-2010). Therefore, the influence of Co and Cu ion leaching was negligible.

Furthermore, as an engineering material, the stability and reusability of $CuCo_2O_4$ is a key factor of practical application. It was studied by using recycled $CuCo_2O_4$ NPs catalysts for ATZ removal under uniform operating conditions (i.e., 150 mg/L $CuCo_2O_4$, 0.2 mM PMS, 5 mg/L ATZ, and initial pH of 6.8). Figure 7 showed that the removal of ATZ gradually decreased from >99% to 97.1%, 96.5%, 90.4%, and 86.7%, in five consecutive experiments. The results show that the $CuCo_2O_4$ NPs had a gradual deactivation in these five recycle experiments because the absorbed intermediates occupied the active sites and small amounts of Co and Cu ions leached into the solution (Ding *et al.* 2017).

Identification of main radicals

Two radical quenching experiments were carried out to identify the main radicals for the removal of ATZ by adding quenching agent into the $CuCo_2O_4$/PMS system. As a quenching agent without α hydrogen, tert-butyl alcohol (TBA) could differentiate $SO_4^{\bullet-}$ and $HO^{\bullet}$ because its reaction rate constant with $HO^{\bullet}$ $(3.8–7.6) \times 10^8\ M^{-1}{\cdot}s^{-1}$ is three orders of magnitude greater than that with $SO_4^{\bullet-}$ ($(4–9.1) \times 10^5\ M^{-1}{\cdot}s^{-1}$) (Li *et al.* 2017). As well, both $SO_4^{\bullet-}$ and $HO^{\bullet}$ could actively react with ethanol (EtOH) which had α hydrogen (Li *et al.* 2017). The reaction rate constant of EtOH and $HO^{\bullet}$ is $(1.2–2.8) \times 10^9\ M^{-1}{\cdot}s^{-1}$ which was almost as large as that of EtOH and $SO_4^{\bullet-}$ ($(1.6–7.7) \times 10^7\ M^{-1}{\cdot}s^{-1}$) (Xu *et al.* 2016). Hence, TBA and EtOH can be applied as two kinds of radical scavengers. With the addition of 50 mM TBA, the removal of ATZ decreased from >99% to

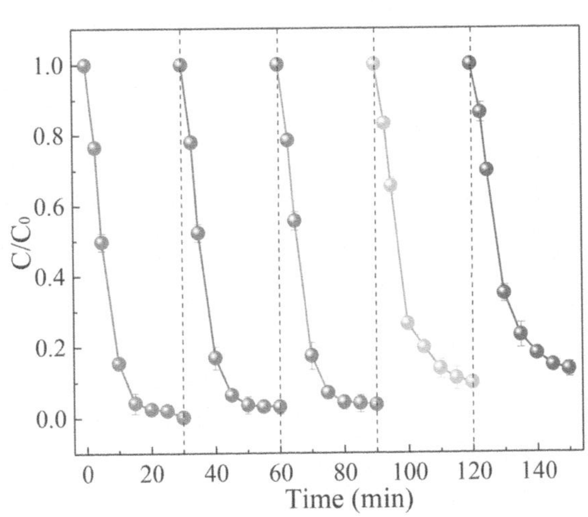

Figure 7 | Reusability of $CuCo_2O_4$ in $CuCo_2O_4$/PMS system for the removal of ATZ. Experiment conditions: $[ATZ]_0 = 5$ mg/L, $[CuCo_2O_4]_0 = 150$ mg/L, $[PMS]_0 = 0.2$ mM, initial pH = 6.8, and reaction time = 30 min.

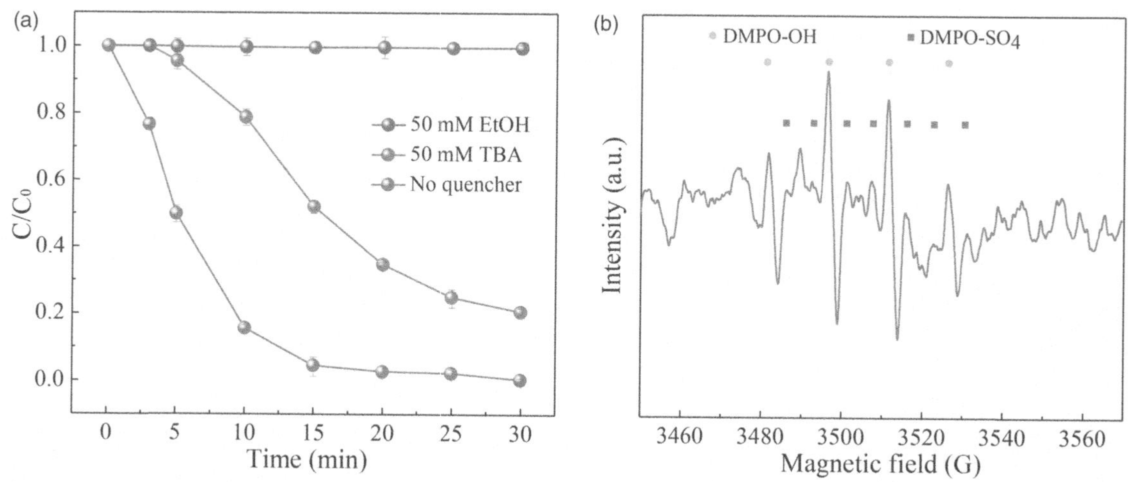

Figure 8 | (a) The radical scavengers on ATZ removal, (b) the EPR spectra of radicals during treatment in $CuCo_2O_4$/PMS system. Experiment conditions: $[ATZ]_0 = 5$ mg/L, $[CuCo_2O_4]_0 = 150$ mg/L, $[PMS]_0 = 0.2$ mM, $[EtOH]_0 = 50$ mM, $[TBA]_0 = 50$ mM, initial pH = 6.8, and reaction time = 30 min.

79.7% in Figure 8(a). The results could prove that $HO^{\cdot}$ made a certain contribution to the removal of ATZ. When 50 mM EtOH was added into aqueous solution, there was almost no ATZ removal because almost all $HO^{\cdot}$ and $SO_4^{\cdot-}$ were quenched by EtOH. It could be concluded from the above results that both $HO^{\cdot}$ and $SO_4^{\cdot-}$ could significantly promote the removal of ATZ and the dominant active radical is $SO_4^{\cdot-}$ in this study.

The EPR spectra were utilized to identify $HO^{\cdot}$ and $SO_4^{\cdot-}$ by using DMPO as the spin trapping agent with the reaction time of 2 min, in order to confirm these two radicals were produced during the activation of PMS. As seen in Figure 8(b), a strong signal of DMPO-$HO^{\cdot}$ with an intensity ratio of 1:2:2:1 was observed, which could confirm the existence of $HO^{\cdot}$. However, the signal of DMPO-$SO_4^{\cdot-}$ was not obvious. The results could be interpreted as follows: DMPO-$SO_4^{\cdot-}$ could be converted to DMPO-$HO^{\cdot}$ via nucleophilic substitution (Equation (14)) (Kang *et al.* 2018), indicating that the produced DMPO-$SO_4^{\cdot-}$ was rapidly consumed (Wang *et al.* 2015). This phenomenon further confirmed that the main radical in the system was $SO_4^{\cdot-}$, which also verified the results of radical quenching experiments.

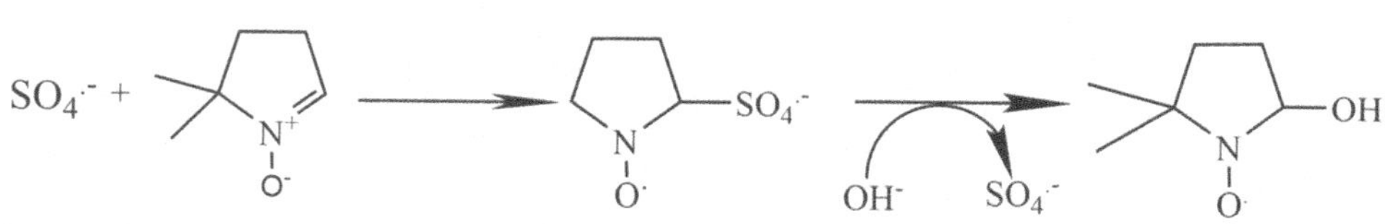

(14)

The possible catalytic mechanism of $CuCo_2O_4$/PMS system

The possible reaction mechanism of the $CuCo_2O_4$/PMS system for ATZ removal is illustrated in Figure 9. Based on the results of Figure 6, a part of ATZ could be directly oxidized by PMS but could not be removed by the adsorption of $CuCo_2O_4$ NPs. Furthermore, the metal ions which played the role of Lewis acid could react with H_2O absorbed on the surface of $CuCo_2O_4$ NPs, producing a large number of active hydroxyls as described in Equations (15) and (16) (Ren *et al.* 2015). This process (Equations (15) and (16)) was the rate-limiting step in the activation of PMS (Ren *et al.* 2015). Next, with the addition of PMS into the reaction solution, HSO_5^- could bind to $CuCo_2O_4$ to form $CuCo_2O_4$-O-H-HSO_5^- for hydrogen bonding. Then, HSO_5^- could react with $\equiv Co^{2+}$-OH^-, to produce $SO_4^{\cdot-}$ and $\equiv CoO^+$ (Equation (17)). The $\equiv CoO^+$ could react with H^+ to form $\equiv Co^{3+}$ (Equation (18)). The cycle of Co was formed after the reaction between $\equiv Co^{3+}$ and HSO_5^- (Equation (19)). Similarly, $\equiv Cu^+$ would also bind to the active hydroxyl to form $\equiv Cu^+$-OH^- (Equation (16)) which could combine with HSO_5^- to produce $SO_4^{\cdot-}$ and $\equiv Cu^{2+}$-OH^- (Equation (20)). Subsequently, $\equiv Cu^{2+}$-OH^- would continue to react with HSO_5^- to form $\equiv Cu^+$ and $SO_5^{\cdot-}$ (Equation (21)). Furthermore, the catalytic activity of $CuCo_2O_4$ could be improved by the reaction of Cu(I) and Co(III) as described in Equation (22), which reflected the strong synergistic effect of Cu and Co (Chen *et al.* 2020). The

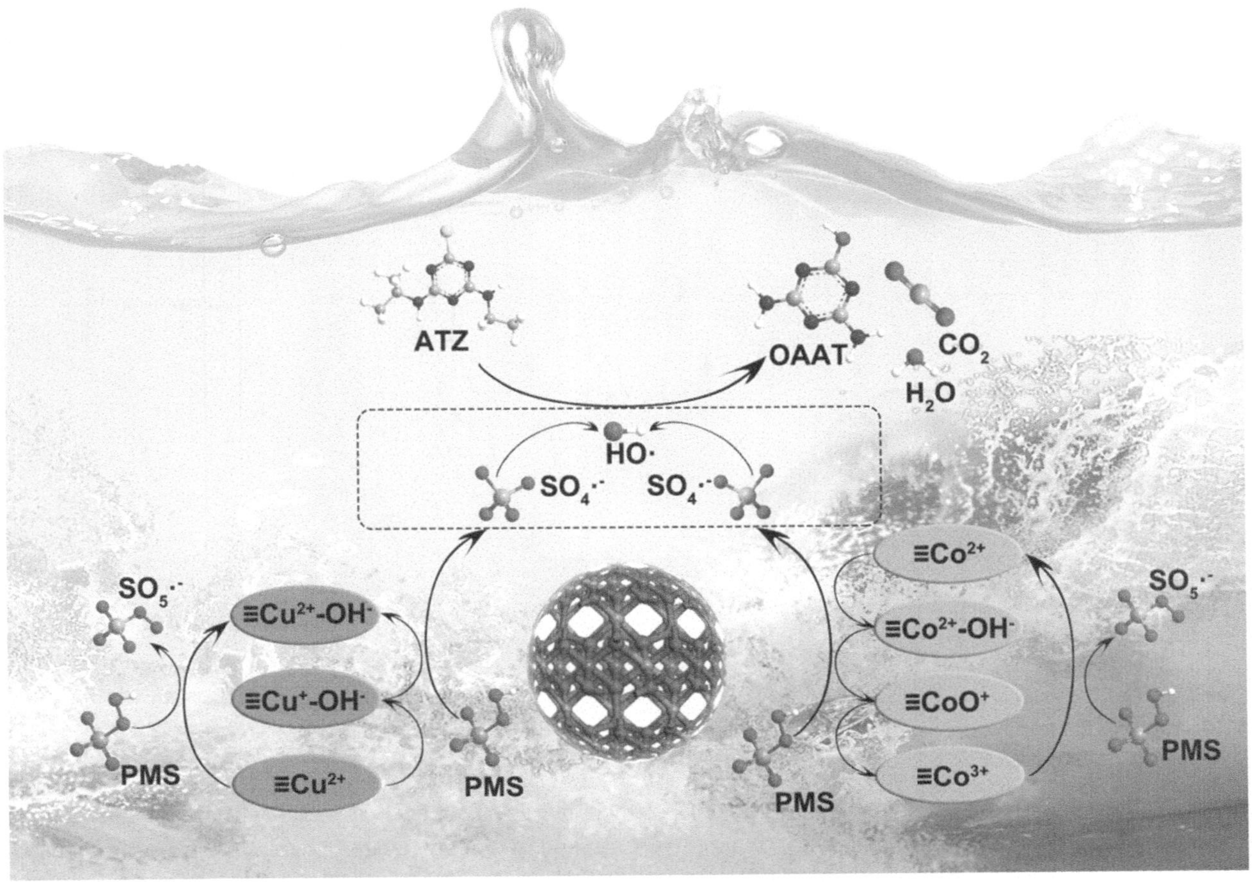

Figure 9 | The possible reaction mechanism of $CuCo_2O_4$/PMS system for the ATZ removal.

HO˙ might be produced as shown in Equations (23) and (24) (Shen *et al.* 2020):

$$\equiv Co^{2+} + H_2O \rightarrow \equiv Co^{2+} - OH^- + H^+ \tag{15}$$

$$\equiv Cu^{+} + H_2O \rightarrow \equiv Cu^{+} - OH^- + H^+ \tag{16}$$

$$\equiv Co^{2+} - OH^- + HSO_5^- \rightarrow \equiv CoO^+ + SO_4^{\bullet -} + H_2O \tag{17}$$

$$\equiv CoO^+ + 2H^+ \rightarrow \equiv Co^{3+} + H_2O \tag{18}$$

$$\equiv Co^{3+} + HSO_5^- \rightarrow \equiv Co^{2+} + H^+ + SO_5^{\bullet -} \tag{19}$$

$$\equiv Cu^{+} - OH^- + HSO_5^- \rightarrow \equiv Cu^{2+} - OH^- + SO_4^{\bullet -} + OH^- \tag{20}$$

$$\equiv Cu^{2+} - OH^- + HSO_5^- \rightarrow \equiv Cu^{+} + H^+ + SO_5^{\bullet -} \tag{21}$$

$$\equiv Co^{3+} + \equiv Cu^{+} \rightarrow \equiv Cu^{2+} + \equiv Co^{2+} \tag{22}$$

$$SO_4^{\bullet -} + H_2O \rightarrow H^+ + HO^{\bullet} + SO_4^{2-} \tag{23}$$

$$SO_4^{\bullet -} + OH^- \rightarrow HO^{\bullet} + SO_4^{2-} \tag{24}$$

Intermediates and possible removal pathways

To speculate on the removal pathway of ATZ, the intermediate products during the reaction process were detected by UPLC-QTOF-MS/MS. The molecular formula, chemical name, abbreviation, charges-to-mass ratio (m/z), the absolute error of m/z, and structural formula of 11 intermediates are listed in Supplementary Material, Table S3. The possible removal pathway of

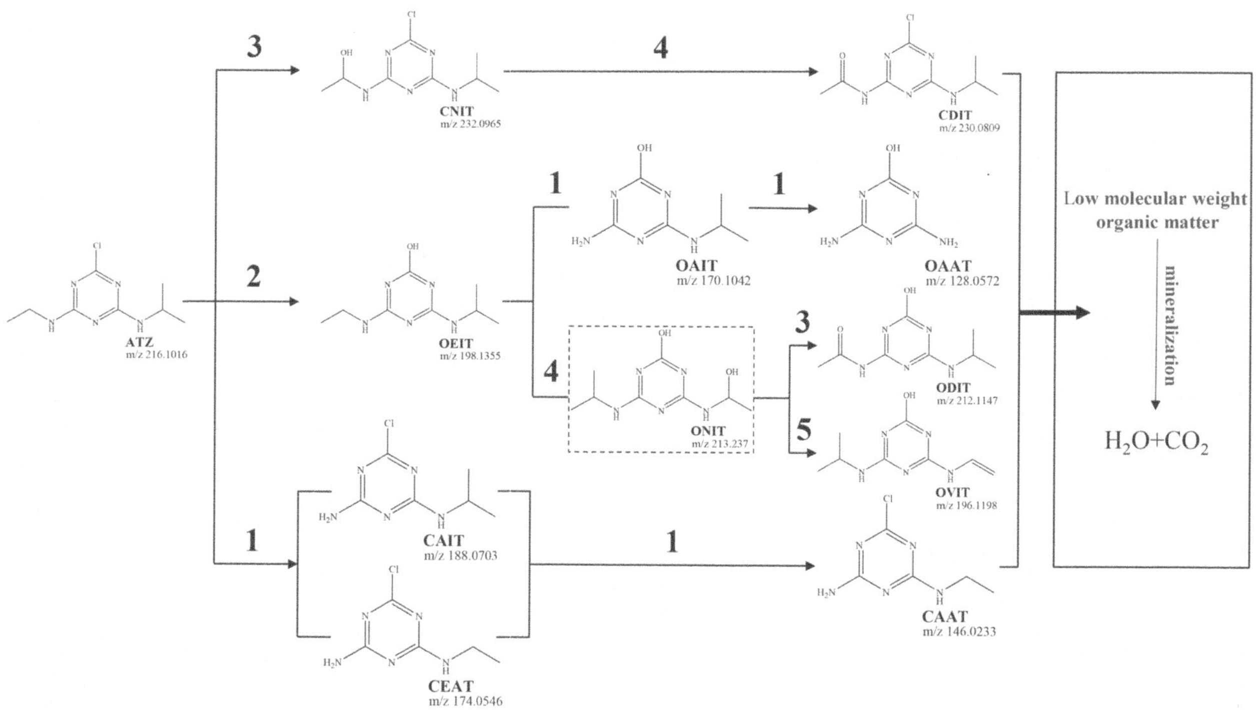

Figure 10 | The possible removal pathways of ATZ in the $CuCo_2O_4$/PMS system.

ATZ is proposed in Figure 10. The removal pathway of ATZ included de-alkylation (path 1), dechlorination–hydroxylation (path 2), alkylic-hydroxylation (path 3), alkylic-oxidation (path 4), and olefination (path 5) as demonstrated in Figure 10.

In path 1, the side chain of ATZ was attacked by $HO^{\bullet}$ and $SO_4^{\bullet-}$, then H in the alkyl group was abstracted to form the carbon-based radical. Subsequently, oxygen would react with the carbon-based radical, resulting in forming the by-product containing peroxide radical, which would bring about the generation of de-alkylation intermediate (e.g., CAIT, CEAT, CAAT, OAIT, and OAAT) (Rejto *et al.* 1983; Acero *et al.* 2000). In path 2, the aromatic ring of ATZ reacted with the $HO^{\bullet}$ and $SO_4^{\bullet-}$ to form radical adduct, which then reacted with water to generate hydroxylated intermediate (e.g., OEIT) (Antoniou *et al.* 2010; Song *et al.* 2012). In path 3, the H on the alkyl group of the side chain could be taken away by free radicals to form carbon-based radicals, which could continue to react with water to produce alkyl-hydroxylation intermediate (e.g., CNIT and ODIT). In path 4, under the action of $HO^{\bullet}$ and $SO_4^{\bullet-}$, the products of path 3 were converted to carbon-based radicals, which reacted with oxygen to generate by-products containing peroxide radical. With water molecular break down, the by-product of the previous step became an alkyl-oxidation product (e.g., ONIT and CDIT). In path 5, alkyl-hydroxylation intermediates were dehydrated to form olefins by-product (e.g., OVIT).

CONCLUSIONS

In this study, the high ATZ removal (>99%) could be attained in the $CuCo_2O_4$/PMS system under the optimal experimental conditions of 150 mg/L $CuCo_2O_4$, 0.2 mM PMS, and initial pH of 6.8. The five co-existing ions (i.e., NO_3^-, HCO_3^-, HPO_4^{2-}, Cl^-, and HA) had some different effects on the removal of ATZ. The effects of these five ions on the ATZ removal were, in descending order, HA > HPO_4^{2-} > HCO_3^- > Cl^- > NO_3^-. Furthermore, compared with PMS alone, $CuCo_2O_4$ alone, CuO/PMS and Co_2O_3/PMS systems, the excellent performance of $CuCo_2O_4$/PMS system was proved. Through five successive runs, on account of the reduction of active sites on the catalyst surface, the ATZ removal decreased from >99% to 97.1%, 96.5%, 90.4%, and 86.7% in turn. In the $CuCo_2O_4$/PMS system, $SO_4^{\bullet-}$ was identified as the dominant radical. In addition, the possible catalytic mechanism of the $CuCo_2O_4$/PMS system was deduced. The removal pathway of ATZ was proposed according to the information of the detected intermediate. In a word, the $CuCo_2O_4$/PMS system is a potentially effective technique for the removal of ATZ.

ACKNOWLEDGEMENTS

The authors would like to acknowledge the financial support from National Natural Science Foundation of China (No. 51878423), Student Innovation and Entrepreneurship Foundation of Sichuan University (No. 201810610418), and Sichuan Province Science and Technology Innovation Seedling Project (No. 2019084).

DATA AVAILABILITY STATEMENT

All relevant data are included in the paper or its Supplementary Information.

REFERENCES

Acero, J. L., Stemmler, K. & Von Gunten, U. 2000 Degradation kinetics of atrazine and its degradation products with ozone and OH radicals: a predictive tool for drinking water treatment. *Environmental Science & Technology* **34**, 591–597.

Anipsitakis, G. P. & Dionysiou, D. D. 2003 Degradation of organic contaminants in water with sulfate radicals generated by the conjunction of peroxymonosulfate with cobalt. *Environmental Science & Technology* **37**, 4790–4797.

Antoniou, M. G., de la Cruz, A. A. & Dionysiou, D. D. 2010 Intermediates and reaction pathways from the degradation of microcystin-LR with sulfate radicals. *Environmental Science & Technology* **44**, 7238–7244.

Buxton, G. V., Greenstock, C. L., Helman, W. P. & Ross, A. B. 1988 Critical-review of rate constants for reactions of hydrated electrons, hydrogen-atoms and hydroxyl radicals ($\cdot OH/\cdot O^-$) in aqueous-solution. *Journal of Physical and Chemical Reference Data* **17**, 513–886.

Chandra, P. N. & Usha, K. 2021 Removal of atrazine herbicide from water by polyelectrolyte multilayer membranes. *Materials Today: Proceedings* **41**, 622–627.

Chen, H., Bramanti, E., Longo, I., Onor, M. & Ferrari, C. 2011 Oxidative decomposition of atrazine in water in the presence of hydrogen peroxide using an innovative microwave photochemical reactor. *Journal of Hazardous Materials* **186**, 1808–1815.

Chen, C., Ma, T., Shang, Y., Gao, B., Jin, B., Dan, H., Li, Q., Yue, Q., Li, Y., Wang, Y. & Xu, X. 2019 In-situ pyrolysis of enteromorpha as carbocatalyst for catalytic removal of organic contaminants: considering the intrinsic N/Fe in enteromorpha and non-radical reaction. *Applied Catalysis B-Environmental* **250**, 382–395.

Chen, C., Liu, L., Li, Y., Li, W., Zhou, L., Lan, Y. & Li, Y. 2020 Insight into heterogeneous catalytic degradation of sulfamethazine by peroxymonosulfate activated with $CuCo_2O_4$ derived from bimetallic oxalate. *Chemical Engineering Journal* **384**, 123257.

De Laat, J., Gallard, H., Ancelin, S. & Legube, B. 1999 Comparative study of the oxidation of atrazine and acetone by H_2O_2/UV, Fe(III)/UV, Fe(III)/H_2O_2/UV and Fe(II) or Fe(III)/H_2O_2. *Chemosphere* **39**, 2693–2706.

Deng, J., Shao, Y., Gao, N., Deng, Y., Zhou, S. & Hu, X. 2013 Thermally activated persulfate (TAP) oxidation of antiepileptic drug carbamazepine in water. *Chemical Engineering Journal* **228**, 765–771.

Ding, D., Liu, C., Ji, Y., Yang, Q., Chen, L., Jiang, C. & Cai, T. 2017 Mechanism insight of degradation of norfloxacin by magnetite nanoparticles activated persulfate: identification of radicals and degradation pathway. *Chemical Engineering Journal* **308**, 330–339.

Ding, W., Huang, X. Y., Zhang, W. D., Wu, F. & Li, J. J. 2019 Sulfite activation by a low-leaching silica-supported copper catalyst for oxidation of As(III) in water at circumneutral pH. *Chemical Engineering Journal* **359**, 1518–1526.

Du, M., Yi, Q., Ji, J., Zhu, Q., Duan, H., Xing, M. & Zhang, J. 2020 Sustainable activation of peroxymonosulfate by the Mo(IV) in mos_2 for the remediation of aromatic organic pollutants. *Chinese Chemical Letters* **31**, 2803–2808.

Feng, M., Qu, R., Zhang, X., Sun, P., Sui, Y., Wang, L. & Wang, Z. 2015a Degradation of flumequine in aqueous solution by persulfate activated with common methods and polyhydroquinone-coated magnetite/multi-walled carbon nanotubes catalysts. *Water Research* **85**, 1–10.

Feng, Y., Liu, J., Wu, D., Zhou, Z., Deng, Y., Zhang, T. & Shih, K. 2015b Efficient degradation of sulfamethazine with $CuCo_2O_4$ spinel nanocatalysts for peroxymonosulfate activation. *Chemical Engineering Journal* **280**, 514–524.

Hayes, T. B., Collins, A., Lee, M., Mendoza, M., Noriega, N., Stuart, A. A. & Vonk, A. 2002 Hermaphroditic, demasculinized frogs after exposure to the herbicide atrazine at low ecologically relevant doses. *Proceedings of the National Academy of Sciences of the United States of America* **99**, 5476–5480.

He, D., Luo, Y., Peng, W., Tang, G., Guo, Q. & Gao, X. 2019 Effects of inorganic ions on the photocatalytic degradation of carbamazepine. *Journal of Water Reuse and Desalination* **9**, 301–309.

Hong, Y., Peng, J., Zhao, X., Yan, Y., Lai, B. & Yao, G. 2019 Efficient degradation of atrazine by CoMgAl layered double oxides catalyzed peroxymonosulfate: optimization, degradation pathways and mechanism. *Chemical Engineering Journal* **370**, 354–363.

Jawad, A., Zhan, K., Wang, H., Shahzad, A., Zeng, Z., Wang, J., Zhou, X., Ullah, H., Chen, Z. & Chen, Z. 2020 Tuning of persulfate activation from a free radical to a nonradical pathway through the incorporation of non-redox magnesium oxide. *Environmental Science and Technology* **54**, 2476–2488.

Ji, Y., Dong, C., Kong, D., Lu, J. & Zhou, Q. 2015 Heat-activated persulfate oxidation of atrazine: implications for remediation of groundwater contaminated by herbicides. *Chemical Engineering Journal* **263**, 45–54.

Jo, Y., Kim, C., Moon, G. H., Lee, J., An, T. & Choi, W. 2018 Activation of peroxymonosulfate on visible light irradiated TiO_2 via a charge transfer complex path. *Chemical Engineering Journal* **346**, 249–257.

Kang, J., Zhou, L., Duan, X., Sun, H. & Wang, S. 2018 Catalytic degradation of antibiotics by metal-free catalysis over nitrogen-doped graphene. *Catalysis Today* **357**, 341–349.

Khaleel, A., Meetani, M., Al Kaabi, N. S., Abdul al, K., Salem, S., Mansour, S., Ahmed, S. H., Bakiro, M., Ramachandran, T., Hamed, F. & Alzamly, A. 2019 Tunable band gap of Bi^{3+}-doped anatase TiO_2 for enhanced photocatalytic removal of Acetaminophen under UV-visible light irradiation. *Journal of Water Reuse and Desalination* **9**, 31–46.

Khan, J. A., He, X., Shah, N. S., Khan, H. M., Hapeshi, E., Fatta-Kassinos, D. & Dionysiou, D. D. 2014 Kinetic and mechanism investigation on the photochemical degradation of atrazine with activated H_2O_2, $S_2O_8^{2-}$ and HSO_5. *Chemical Engineering Journal* **252**, 393–403.

Kibsgaard, J., Chen, Z., Reinecke, B. N. & Jaramillo, T. F. 2012 Engineering the surface structure of mos_2 to preferentially expose active edge sites for electrocatalysis. *Nature Materials* **11**, 963–969.

Klamerth, N., Miranda, N., Malato, S., Agueera, A., Fernandez-Alba, A. R., Maldonado, M. I. & Coronado, J. M. 2009 Degradation of emerging contaminants at low concentrations in MWTPs effluents with mild solar photo-Fenton and TiO_2. *Catalysis Today* **144**, 124–130.

Kuosa, M., Laari, A., Solonen, A., Haario, H. & Kallas, J. 2009 Multicomponent reaction kinetics for the ozonation of p-nitrophenol and its decomposition products under acidic conditions at constant pH. *Chemical Engineering Science* **64**, 2332–2342.

Lai, L., Zhou, H. & Lai, B. 2018 Heterogeneous degradation of bisphenol A by peroxymonosulfate activated with vanadium-titanium magnetite: performance, transformation pathways and mechanism. *Chemical Engineering Journal* **349**, 633–645.

Lee, J., von Gunten, U. & Kim, J.-H. 2020 Persulfate-based advanced oxidation: critical assessment of opportunities and roadblocks. *Environmental Science & Technology* **54**, 3064–3081.

Li, S.-X., Wei, D., Mak, N.-K., Cai, Z., Xu, X.-R., Li, H.-B. & Jiang, Y. 2009 Degradation of diphenylamine by persulfate: performance optimization, kinetics and mechanism. *Journal of Hazardous Materials* **164**, 26–31.

Li, J., Ren, Y., Ji, F. & Lai, B. 2017 Heterogeneous catalytic oxidation for the degradation of p-nitrophenol in aqueous solution by persulfate activated with $CuFe_2O_4$ magnetic nano-particles. *Chemical Engineering Journal* **324**, 63–73.

Li, J., Xu, M. J., Yao, G. & Lai, B. 2018 Enhancement of the degradation of atrazine through $CoFe_2O_4$ activated peroxymonosulfate (PMS) process: kinetic, degradation intermediates, and toxicity evaluation. *Chemical Engineering Journal* **348**, 1012–1024.

Lifka, J., Ondruschka, B. & Hofmann, J. 2003 The use of ultrasound for the degradation of pollutants in water: aquasonolysis – a review. *Engineering in Life Sciences* **3**, 253–262.

Liu, W., Feng, Y., Sun, L., Zhang, Y., Wang, G., Zhao, L., Meng, M., Li, J. & Liu, K. 2018 Hierarchical $CuCo_2O_4$ nanourchin supported by Ni foam with superior electrochemical performance. *Journal of Alloys and Compounds* **756**, 68–75.

Lou, X., Wu, L., Guo, Y., Chen, C., Wang, Z., Xiao, D., Fang, C., Liu, J., Zhao, J. & Lu, S. 2014 Peroxymonosulfate activation by phosphate anion for organics degradation in water. *Chemosphere* **117**, 582–585.

Luo, C., Ma, J., Jiang, J., Liu, Y., Song, Y., Yang, Y., Guan, Y. & Wu, D. 2015 Simulation and comparative study on the oxidation kinetics of atrazine by UV/H_2O_2, UV/HSO_5^- and $UV/S_2O_8^{2-}$. *Water Research* **80**, 99–108.

Lutze, H. V., Kerlin, N. & Schmidt, T. C. 2015 Sulfate radical-based water treatment in presence of chloride: formation of chlorate, inter-conversion of sulfate radicals into hydroxyl radicals and influence of bicarbonate. *Water Research* **72**, 349–360.

Mahamuni, N. N. & Adewuyi, Y. G. 2010 Advanced oxidation processes (AOPs) involving ultrasound for waste water treatment: a review with emphasis on cost estimation. *Ultrasonics Sonochemistry* **17**, 990–1003.

Neta, P., Huie, R. E. & Ross, A. B. 1988 Rate constants for reactions of inorganic radicals in aqueous-solution. *Journal of Physical and Chemical Reference Data* **17**, 1027–1284.

Nimai, S., Zhang, H., Wu, Z., Li, N. & Lai, B. 2020 Efficient degradation of sulfamethoxazole by acetylene black activated peroxydisulfate. *Chinese Chemical Letters* **31**, 2657–2660.

Olmez-Hanci, T., Imren, C., Kabdasli, I., Tunay, O. & Arslan-Alaton, I. 2011 Application of the UV-C photo-assisted peroxymonosulfate oxidation for the mineralization of dimethyl phthalate in aqueous solutions. *Photochemical & Photobiological Sciences* **10**, 408–413.

Panayotov, D. & Mehandjiev, D. 1987 Surface-state and activity of $CuCo_2O_4$ during the reduction of nitric-oxide with carbon-monoxide. *Applied Catalysis* **34**, 65–76.

Pelizzetti, E., Maurino, V., Minero, C., Carlin, V., Pramauro, E., Zerbinati, O. & Tosato, M. L. 1990 Photocatalytic degradation of atrazine and other s-triazine herbicides. *Environmental Science & Technology* **24**, 1559–1565.

Rastogi, A., Ai-Abed, S. R. & Dionysiou, D. D. 2009 Sulfate radical-based ferrous-peroxymonosulfate oxidative system for PCBs degradation in aqueous and sediment systems. *Applied Catalysis B-Environmental* **85**, 171–179.

Rejto, M., Saltzman, S., Acher, A. J. & Muszkat, L. 1983 Identification of sensitized photo-oxidation products of s-triazine herbicides in water. *Journal of Agricultural and Food Chemistry* **31**, 138–142.

Ren, Y., Lin, L., Ma, J., Yang, J., Feng, J. & Fan, Z. 2015 Sulfate radicals induced from peroxymonosulfate by magnetic ferrospinel MFe_2O_4 ($M = Co$, Cu, Mn, and Zn) as heterogeneous catalysts in the water. *Applied Catalysis B-Environmental* **165**, 572–578.

Shahzad, A., Jawad, A., Ifthikar, J., Chen, Z. & Chen, Z. 2019 The hetero-assembly of reduced graphene oxide and hydroxide nanosheets as superlattice materials in PMS activation. *Carbon* **155**, 740–755.

Sharma, J., Mishra, I. M., Dionysiou, D. D. & Kumar, V. 2015 Oxidative removal of bisphenol A by UV-C/peroxymonosulfate (PMS): kinetics, influence of co-existing chemicals and degradation pathway. *Chemical Engineering Journal* **276**, 193–204.

Shen, Z., Zhou, H., Pan, Z., Guo, Y., Yuan, Y., Yao, G. & Lai, B. 2020 Degradation of atrazine by Bi_2MoO_6 activated peroxymonosulfate under visible light irradiation. *Journal of Hazardous Materials* **400**, 123187.

Song, W., Yan, S., Cooper, W. J., Dionysiou, D. D. & O'Shea, K. E. 2012 Hydroxyl radical oxidation of cylindrospermopsin (cyanobacterial toxin) and its role in the photochemical transformation. *Environmental Science & Technology* **46**, 12608–12615.

Song, D., Li, T., Wei, Y. Y. & Xu, Z. R. 2020 Controlled formation of porous $CuCo_2O_4$ nanorods with enhanced oxidase and catalase catalytic activities using bimetal-organic frameworks as templates. *Colloids and Surface B: Biointerfaces* **188**, 110764.

Wang, Y., Indrawirawan, S., Duan, X., Sun, H., Ang, H. M., Tadé, M. O. & Wang, S. 2015 New insights into heterogeneous generation and evolution processes of sulfate radicals for phenol degradation over one-dimensional α-MnO_2 nanostructures. *Chemical Engineering Journal* **266**, 12–20.

Wang, A., Hu, X., Wan, Y., Mahai, G., Jiang, Y., Huo, W., Zhao, X., Liang, G., He, Z., Xia, W. & Xu, S. 2020a A nationwide study of the occurrence and distribution of atrazine and its degradates in tap water and groundwater in China: assessment of human exposure potential. *Chemosphere* **252**, 126533.

Wang, X., Pu, X., Yuan, Y., Xiang, Y., Zhang, Y., Xiong, Z., Yao, G. & Lai, B. 2020b An old story with new insight into the structural transformation and radical production of micron-scale zero-valent iron on successive reactivities. *Chinese Chemical Letters* **31**, 2634–2640.

Wang, Y. P., Ji, H. D., Liu, W., Xue, T. S., Liu, C., Zhang, Y. T., Liu, L. Y., Wang, Q., Qi, F., Xu, B. B., Tsang, D. C. W. & Chu, W. 2020c Novel $CuCo_2O_4$ composite spinel with a meso-macroporous nanosheet structure for sulfate radical formation and benzophenone-4 degradation: interface reaction, degradation pathway, and DFT calculation. *ACS Applied Materials & Interfaces* **12**, 20522–20535.

Xu, Y., Ai, J. & Zhang, H. 2016 The mechanism of degradation of bisphenol A using the magnetically separable $CuFe_2O_4$/peroxymonosulfate heterogeneous oxidation process. *Journal of Hazardous Materials* **309**, 87–96.

Xu, H., Wang, D., Ma, J., Zhang, T., Lu, X. & Chen, Z. 2018 A superior active and stable spinel sulfide for catalytic peroxymonosulfate oxidation of bisphenol S. *Applied Catalysis B: Environmental* **238**, 557–567.

Xu, K., Ma, S., Shen, Y., Ren, Q., Yang, J., Chen, X. & Hu, J. 2019 $CuCo_2O_4$ nanowire arrays wrapped in metal oxide nanosheets as hierarchical multicomponent electrodes for supercapacitors. *Chemical Engineering Journal* **369**, 363–369.

Zhang, Y. & Zhou, M. 2019 A critical review of the application of chelating agents to enable Fenton and 1 Fenton-like reactions at high pH values. *Journal of Hazardous Materials* **362**, 436–450.

Zhang, B.-T., Zhang, Y., Teng, Y. & Fan, M. 2015 Sulfate radical and its application in decontamination technologies. *Critical Reviews in Environmental Science and Technology* **45**, 1756–1800.

Zhang, H., Ji, Q., Lai, L., Yao, G. & Lai, B. 2019a Degradation of p-nitrophenol (PNP) in aqueous solution by mFe/Cu-air-PS system. *Chinese Chemical Letters* **30**, 1129–1132.

Zhang, R., Wan, Y., Peng, J., Yao, G., Zhang, Y. & Lai, B. 2019b Efficient degradation of atrazine by $LaCoO_3/Al_2O_3$ catalyzed peroxymonosulfate: performance, degradation intermediates and mechanism. *Chemical Engineering Journal* **372**, 796–808.

Zhao, L., Ji, Y., Kong, D., Lu, J., Zhou, Q. & Yin, X. 2016a Simultaneous removal of bisphenol A and phosphate in zero-valent iron activated persulfate oxidation process. *Chemical Engineering Journal* **303**, 458–466.

Zhao, S., Li, Q., Li, F. & Liang, Z. 2016b Synthesis of spinel $CuCo_2O_4$ nanoparticles and its application in p-nitrophenol reduction. *Journal of Sol-Gel Science and Technology* **81**, 544–555.

Zhou, X. J., Shi, P. H., Qin, Y. F., Fan, J. C., Min, Y. L. & Yao, W. F. 2016 Synthesis of Co_3O_4/graphene composite catalysts through CTAB-assisted method for orange II degradation by activation of peroxymonosulfate. *Journal of Materials Science-Materials in Electronics* **27**, 1020–1030.

Zou, X., Zhou, T., Mao, J. & Wu, X. 2014 Synergistic degradation of antibiotic sulfadiazine in a heterogeneous ultrasound-enhanced Fe^0/persulfate Fenton-like system. *Chemical Engineering Journal* **257**, 36–44.

First received 2 June 2021; accepted in revised form 19 July 2021. Available online 3 November 2021

Water Recycling and Low-Carbon Development doi: 10.2166/wrd.2021.053

Kinetics of diatrizoate degradation by ozone and the formation of disinfection by-products in the sequential chlorination

Chen-Yan Hu[a,b], Si-Cheng Ren[a], Yi-Li Lin[c,*], Ji-Chen Zhang[a], Ye-Ye Zhu[a], Cun Xiong[a] and Qiang-Bin Wang[a]

[a] College of Environmental and Chemical Engineering, Shanghai Engineering Research Center of Energy-Saving in Heat Exchange Systems, Shanghai University of Electric Power, Shanghai 200090, China

[b] Shanghai Institute of Pollution Control and Ecological Security, Shanghai 200092, China

[c] Department of Safety, Health and Environmental Engineering, National Kaohsiung University of Science and Technology, Kaohsiung 824, Taiwan, ROC

*Corresponding author. E-mail: yililin@nkust.edu.tw

Y-LL, 0000-0003-3426-3565; J-CZ, 0000-0002-1930-8174

ABSTRACT

In this study, we studied the degradation kinetics of a common iodine contrast agent, diatrizoate, by ozone and the formation of disinfection by-products (DBPs) in the sequential chlorination. Effects of ozone concentration, solution pH, and bromide concentration on diatrizoate degradation were evaluated. The results indicate that diatrizoate can be effectively degraded (over 80% within 1 h) by ozone, and the degradation kinetics can be well described using the pseudo-first-order kinetic model. The pseudo-first-order rate constant (k_{obs}) of diatrizoate degradation significantly increased with increasing ozone concentration and decreasing bromide concentration. The k_{obs} kept increasing with the increase of pH value and reached a maximum of 6.5 ($\pm$0.05) $\times$ 10^{-2} min^{-1} at pH 9. As the ozone concentration gradually increased from 0.342 to 1.316 mg/L, the corresponding k_{obs} of diatrizoate degradation increased from 1.76 ($\pm$0.20) $\times$ 10^{-3} to 4.22 ($\pm$0.3) $\times$ 10^{-2} min^{-1}. The bromide concentration exhibited a strong inhibitory effect on diatrizoate degradation because of the competition for ozone with diatrizoate. Trichloromethane was the only detected DBP in the subsequent chlorination in the absence of bromide. However, in the presence of bromide, six other DBPs were detected, and bromochloroiodomethane and tribromomethane became the major products with concentrations 1–2 orders higher than those of the other DBPs. In order to provide safe drinking water to the public, water should be maintained at circumneutral pH values and low bromine concentrations ($<$5 μM) before reaching the chlorine disinfection process to effectively control the formation of DBPs.

Key words: bromide, chlorine, disinfection by-product (DBP), iodinated X-ray contrast media (ICM), kinetics, ozone

HIGHLIGHTS

- Diatrizoate ozonation followed pseudo-first-order kinetics.
- The diatrizoate degradation rate can be significantly increased at pH $>$ 7.
- •OH played a major role in diatrizoate degradation during ozonation.
- Bromide can strongly inhibit diatrizoate degradation because of the competition for ozone with diatrizoate.
- The formation of DBPs can be controlled at circumneutral pH and low bromine concentrations ($<$5 μM).

GRAPHICAL ABSTRACT

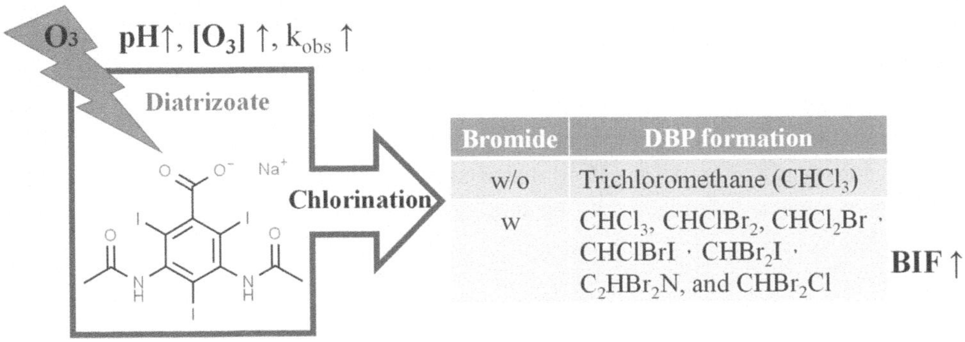

1. INTRODUCTION

Iodinated X-ray contrast media (ICM) are a class of pharmaceutical products widely used to increase the contrast of tissue, structures, and details of organs and blood vessels during medical imaging at hospitals and medical centers (Hu *et al.* 2019a). The annual global consumption of ICM is approximately 3.5×10^6 kg (Perez & Barcelo 2007; Wang *et al.* 2016), and the annual estimated production of ICM is more than 5×10^6 kg (Gharekhanloo & Torabian 2012; Hu *et al.* 2019a). Typically, 95% of non-metabolized ICM can be eliminated through urine and feces within 24 h after adoption (Sandra Perez *et al.* 2006). The molecular structures of ICM consist of 2,4,6-triiodinated benzoic derivatives, and their molecular weights vary between 600 and 900 Da (Jeong *et al.* 2017). The structures of ICM are very stable and cannot be effectively removed by traditional water treatment processes (Hu *et al.* 2020a). Therefore, trace amounts of ICM are commonly detected in surface waters (Duirk *et al.* 2011; Kormos *et al.* 2011). In a previous study, many kinds of ICM were detected in the rivers (Duirk *et al.* 2011) of drinking water sources in the US at the levels of 10–2,700 ng/L, including iopamidol, iomeprol, iopromide, iohexol, and diatrizoate.

ICM can be divided into ionic and nonionic groups (Hu *et al.* 2019b), which are all commonly used ICM nowadays (Hu *et al.* 2020b). Diatrizoate is a nonionic ICM that is especially resistant to conventional wastewater and drinking water treatment processes (Sugihara *et al.* 2013). Therefore, many researchers have focused on the degradation of diatrizoate using advanced oxidation techniques (Hennebel *et al.* 2010; Velo-Gala *et al.* 2014; Azerrad *et al.* 2016; Polo *et al.* 2016; Meng *et al.* 2017). For example, Hennebel *et al.* (2010) removed diatrizoate using a catalytically active membrane, and the removal efficiency reached 77% after 48 h. Velo-Gala *et al.* (2014) compared the degradation of diatrizoate in a solution with various advanced oxidation processes (AOPs) using iron salts and ultraviolet light (UV) radiation. Experimental results showed that $UV/K_2S_2O_8$ is more effective than UV/H_2O_2 with a higher reaction rate constant for diatrizoate degradation. Polo *et al.* used solar radiation to degrade diatrizoate (Polo *et al.* 2016) and reported that the formation of free radical promoting substances (including complexes of peroxygen and iron hydroperoxide) can absorb radiation in the solar radiation area, resulting in 100% sunlight degradation of diatrizoate. A German municipal sewage treatment plant (STP) using ozonation only achieved 14% removal of diatrizoate (Ternes *et al.* 2003).

Halogenated disinfection by-products (DBPs) are produced by natural organic matter (NOM) reacting with chlorine during drinking water disinfection (Jiang *et al.* 2020). The presence of ICM in the source water may also react with chlorine to produce highly toxic iodinated DBPs (I-DBPs), which have aroused high concerns for public health (Jeong *et al.* 2017). DBPs have chronic cytotoxicity and genomic induction. Research results indicated that 13 kinds of haloacetamides can damage the DNA of Chinese hamster ovary cells (Michael *et al.* 2008). Moreover, Jiang *et al.* (2017) reported that the presence of bromide ions in the raw water may participate in the reactions between NOM and chlorine to generate brominated-DBPs (Br-DBPs), which are more carcinogenic and mutagenic than their chlorinated analogues. Epidemiological studies have consistently observed a correlation between chlorine consumption in drinking water and increasing risk of bladder cancer (Li & Mitch 2018). DBPs and their toxic effects will put pressure on the living creatures in the environment and organisms (Shang *et al.* 2019; Ranjan *et al.* 2021). ICM is a potential source of toxic I-DBPs during water disinfection, such as iopamidol (Mao *et al.* 2020). It has been reported that I-DBPs are generated during chlorine disinfection of iodide-containing water in sewage treatment systems (Allard *et al.* 2013; Xia *et al.* 2017; Liu *et al.* 2018; Zhang *et al.* 2019). Allard *et al.* (2013) reported that in iodide-containing waters, I-DBPs can be produced during chlorination or especially chloramination; a pre-ozonation step to oxidize iodide to iodate is an efficient process to mitigate I-DBP formation. I-DBPs have recently become an emerging group of DBPs with public health concerns (Yolanta *et al.* 2014). To the authors' best knowledge, no literature regarding the degradation kinetics of diatrizoate ozonation and DBP formation after post-chlorination have been reported, which are worth research attention in order to provide safe drinking water to the public.

Therefore, the purposes of this study were (1) to investigate the kinetics of diatrizoate degradation during ozone oxidation, (2) to evaluate the effects of bromide concentration and solution pH on diatrizoate degradation, and (3) to investigate the formation of regulated and emerging DBPs of ozonated diatrizoate in the subsequent chlorination to simulate water distribution in the pipelines.

2. MATERIALS AND METHODS

2.1. Chemicals and reagents

Diatrizoate (>99.0%) was obtained from Sigma-Aldrich (St. Louis, MO, USA). Sodium bromide (NaBr), potassium iodide (KI), sodium bicarbonate ($NaHCO_3$), nitrobenzene ($C_6H_5NO_2$), tert-butanol (C_4H_9OH, TBA), sodium carbonate (Na_2CO_3), potassium dihydrogen phosphate (KH_2PO_4), sodium thiosulfate ($Na_2S_2O_3$), methanol (CH_4O), sodium hydroxyl (NaOH), and hydrochloric acid (HCl) were acquired from Sinopharm Chemical Reagent Co., Ltd (Shanghai, China). US EPA 551A and 551B halogenated DBP standard solutions, including trihalomethanes (THMs) and haloacetonitriles, were purchased from Sigma-Aldrich (USA). Other standard solutions of iodinated trihalomethanes (I-THMs), including CHI_3, $CHBr_2I$, CHBrClI, $CHBrI_2$, $CHCl_2I$, and $CHClI_2$ were purchased from CanSyn Chemical Corporation (Toronto, ON, Canada). Methyl tert-butyl ether (MtBE) was purchased from J.T. Baker (USA). Ultrapure water produced by a Milli-Q water purification system (Millipore, USA) was used in the experiments to prepare the solutions. All chemicals were at least at analytical purity, and all solvents are at high-performance liquid chromatograph (HPLC) grade and used without further purification.

2.2. Ozone reaction system

The schematic diagram of the ozone reaction system is displayed in Figure 1. Ozone gas was produced by a COM-AD-01 ozone generator (Anseros, Germany). The desired ozone concentration was controlled by adjusting the flow rate of a stream of pure ozone gas into the reactor. Before each experiment, the concentration of dissolved ozone in the reactor was monitored by measuring the amount of I_2 formed in the reaction between O_3 and potassium iodide solution according to the sodium thiosulfate titration method based on the following equations (Giuseppe *et al.* 2007).

$$O_3 + 2KI + H_2O \rightarrow O_2 + I_2 + 2KOH \tag{1}$$

$$I_2 + 2Na_2S_2O_3 \rightarrow 2NaI + Na_2S_4O_6 \tag{2}$$

An internal micro pump (MP-10R, Keyuan Pumps Co., Ltd, Shanghai, China) and an external peristaltic pump (DP-130, Shanghai Magnetic Pump Co., Ltd, Shanghai, China) were adopted to circulate the water inside the double-layer glass reactor and the water flowed in the outer layer, respectively, to maintain the system temperature at 25 ± 1 °C.

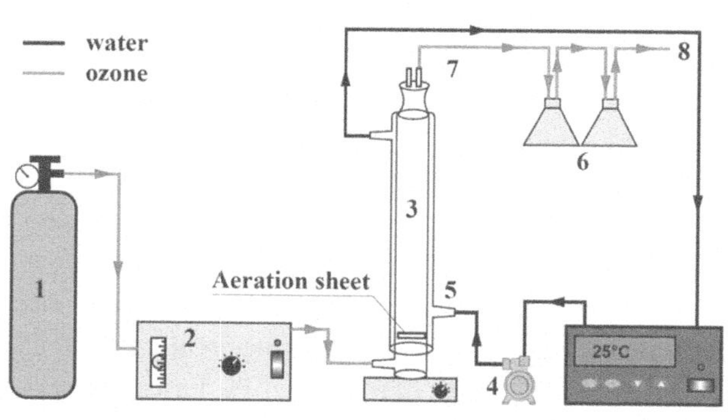

Figure 1 | Schematic diagram of the experimental setup of the ozonation system. 1. High purity oxygen cylinder, 2. Ozone generator, 3. Double-layer glass reactor, 4. Circulating pump, 5. Water bath, 6. Tail gas absorption bottle, 7. Drain outlet, 8. Sampling port.

2.3. Experimental procedures

The ozonation experiments were carried out in a ventilated hood. Diatrizoate stock solution (1 mM) was prepared to make each reacting solution of 20 μM and 200 mL using ultrapure water. The pH value of the reaction solution was adjusted to a desired value (pH 5–9) by adding a small amount (<5 mL) of H_2SO_4 (0.18M) or NaOH (1M) in the presence of 10 mM phosphate buffer. In order to ensure the purity of the produced ozone, the ozone gas produced by the generator was pre-ventilated for 30 min into a saturated potassium iodide solution to adsorb ozone before releasing into the atmosphere. After pre-aeration, the pure ozone gas was passed into the double-layer reactor to start the reaction. For the kinetic experiments, the ozone concentrations were set at 0.342–1.316 mg/L, while for the degradation experiments, the O_3 concentration was fixed at 0.713 mg/L. At a certain time interval, 1 mL of the reaction solution was taken from the sampling port and transferred into an HPLC vial containing $Na_2S_2O_3$ (concentration more than 1.2 times the ozone concentration) to quench the reaction, and the sample was stored at 4 °C in the dark for analyzing the residual diatrizoate concentration using an HPLC as soon as possible. Parallel experiments were performed simultaneously. On the other hand, TBA with concentration more than 1.2 times the ozone concentration was dosed in the solution to verify the contribution of hydroxyl radicals (•OH) on diatrizoate degradation during ozonation.

To study the effect of bromide concentration on diatrizoate degradation, the experiments were performed by dosing bromide concentrations of 0–150 μM at pH 7 and ozone concentration of 0.713 mg/L. For the evaluation of DBP formation after diatrizoate ozonation in the sequential chlorination to simulate drinking water distribution in the pipelines, the diatrizoate solution was prepared in a 45 mL vial dosing 100 μM free chlorine, sealed and preserved in a thermostatic incubator at 25 °C for 3 d. Then, all samples were quenched with $Na_2S_2O_3$. Each DBP formation experiment was performed in duplicate with averaged values being reported in this study.

2.4. Analytical methods

An Agilent 1200 infinity series HPLC system (Agilent, USA) was used to analyze the concentration of diatrizoate with an XTerra MS C18 column (5 μm, 250 mm × 4.6 mm, Waters, USA), a UV detector, and an autosampler with an injection volume of 5 μL. The mobile phase was composed of 10%/90% (v/v) methanol and 1% phosphoric acid solution (v/v) at a flow rate of 1 mL min^{-1}, and the volume ratio is 10%/90%. The detection wavelength of diatrizoate was 237 nm.

A GC (GC-2010 Plus, Shimadzu, Japan) equipped with an RTX-5 fused-silica capillary column (30 m × 0.25 mm id, 0.25 μm film thickness) and an electron capture detector was used to analyze the concentrations of the formed DBPs. The analysis was conducted according to the US EPA method 551.1 (Richardson *et al.* 2007).

3. RESULTS AND DISCUSSION

3.1. Kinetics of diatrizoate degradation during ozone oxidation

Figure 2 displays the pseudo-first-order kinetics plot of diatrizoate degradation during ozonation with different initial ozone concentrations. Diatrizoate can be degraded by more than 91% after 60 min ozonation. The fitted line in Figure 2 exhibited a high linearity ($R^2 > 0.99$), indicating that diatrizoate degradation follows pseudo-first-order kinetics during ozonation. The degradation rate of diatrizoate can be expressed as follows:

$$\frac{-d[\text{Diatrizoate}]}{dt} = k_{\text{obs}}[\text{Diatrizoate}] \tag{3}$$

where [Diatrizoate] represents the diatrizoate concentration at time t, and k_{obs} (min^{-1}) represents the observed pseudo-first-order rate constant of diatrizoate degradation during ozonation.

The slope in Figure 2 is k_{obs}. As the initial ozone concentration increased from 0.342 to 1.316 mg/L, k_{obs} increased linearly from 1.76 ($\pm$0.20) $\times 10^{-3}$ to 4.22 ($\pm$0.3) $\times 10^{-2}$ min^{-1} ($R^2 > 0.923$), which can be explained by the increasing concentrations of dissolved ozone and the resulted hydroxyl radicals in the solution (Huang *et al.* 2020a). Therefore, the ozonation degradation of diatrizoate can be further expressed as follows:

$$\frac{-d[\text{Diatrizoate}]}{dt} = k_{\text{obs}}[\text{Diatrizoate}] = k_{\text{app}}[O_3][\text{Diatrizoate}] \tag{4}$$

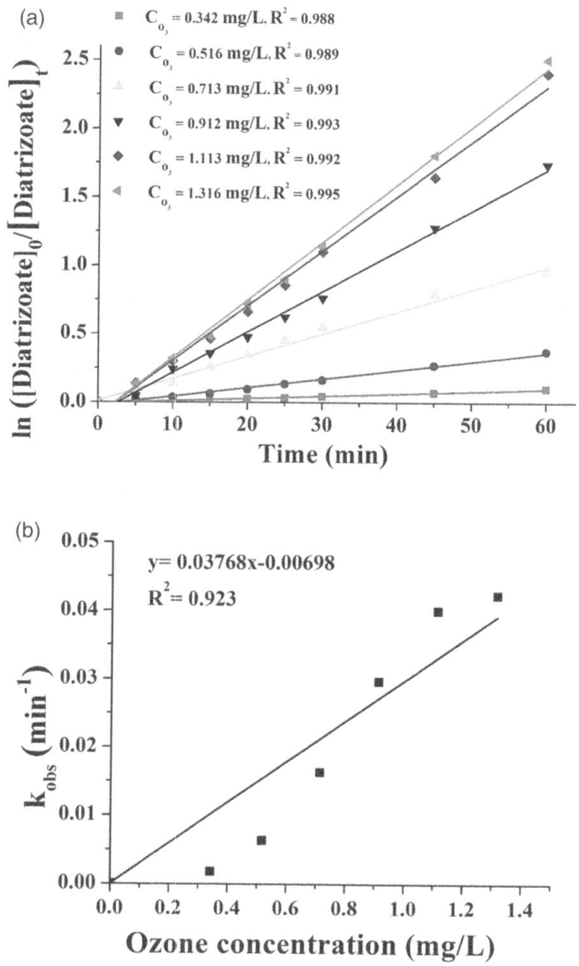

Figure 2 | (a) Pseudo-first-order kinetics plot of diatrizoate degradation at different ozone concentrations, [diatrizoate]0 = 20 μM, [phosphate buffer]0 = 10 mM, and temperature = 25 °C. (b) Correlation between k_{obs} and ozone concentration.

where k_{app} represents the apparent second-order reaction rate constant of the entire reaction. In sum, the ozonation degradation of diatrizoate fitted the second-order kinetics well, one order in ozone concentration and one order in diatrizoate concentration. The relationship between k_{obs} and k_{app} can be expressed (Equation (5)) and calculated as 0.03768 L (min mg)$^{-1}$, which is higher than that of iohexol degradation during ozonation (Hu *et al.* 2020a).

$$k_{app} = \frac{k_{obs}}{[O_3]_T} = 0.03768 \tag{5}$$

3.2. Effect of tert-butanol (TBA) on diatrizoate degradation during ozonation

In order to verify the contribution of hydroxyl radicals (•OH) on diatrizoate degradation during ozonation, the commonly used quencher, TBA, was dosed during the reaction. As displayed in Figure 3(a) and 3(b), the degradation rate of diatrizoate decreased sharply in the absence of TBA, while the degradation rate of diatrizoate only slightly decreased in the presence of TBA. Therefore, it can be concluded that •OH played the major role in diatrizoate degradation during ozonation along with partial contribution of ozone and other radicals.

3.3. Effect of solution pH on the degradation of diatrizoate during ozonation

The effect of solution pH on diatrizoate degradation during ozonation was investigated at pH 5.5–9, and the results are presented in Figure 4. The degradation rate of diatrizoate gradually increased from 4.17 ($\pm$0.04) $\times$ 10^{-3} min^{-1} to the maximum

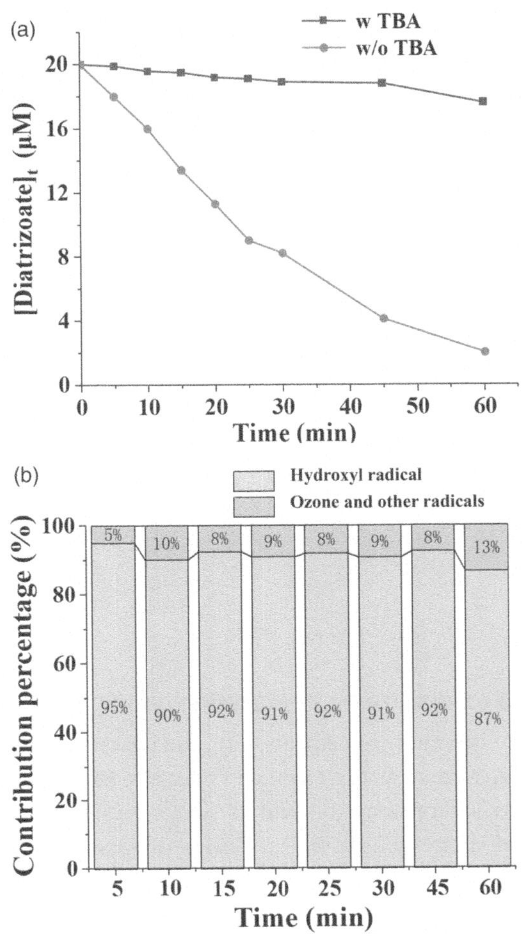

Figure 3 | (a) Effect of tert-butanol on the degradation of diatrizoate during ozonation, [Ozone] = 0.713 mg/L, [diatrizoate]0 = 20 μM, [phosphate buffer]0 = 10 mM, [TBA] = 0.856 mg/L, and temperature = 25 °C. (b) Contribution percentage of different radicals to diatrizoate degradation.

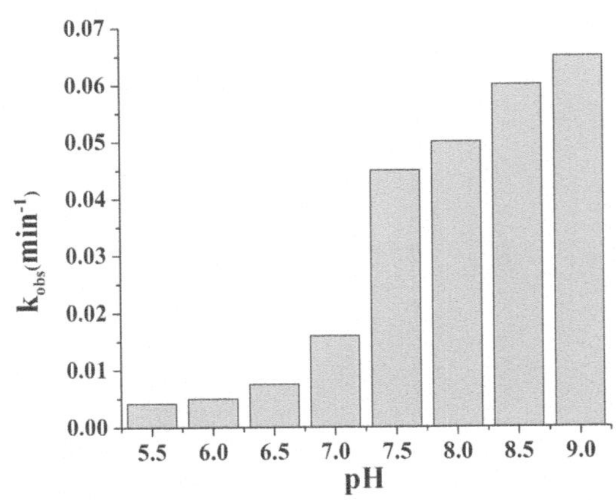

Figure 4 | Effect of solution pH on the rate constant of diatrizoate degradation during ozonation, [Ozone] = 0.713 mg/L, [diatrizoate]0 = 20 μM, [phosphate buffer] 0 = 10 mM, and temperature = 25 °C.

of 6.5 ($\pm$0.05) $\times$ 10^{-2} min^{-1} as pH increased from 5.5 to 9.0. The dosed diatrizoate can be completely decomposed at pH 7.5–9.0 within an hour. The rate constants of diatrizoate degradation during ozonation in alkaline conditions were significantly higher than those in acidic conditions. This phenomenon can be explained by the major contribution of hydroxyl radicals

(•OH) to diatrizoate degradation, which will be further discussed in Section 3.5. At pH > 7, increasing OH^- concentration in the solution can promote the formation of hydroxyl radicals according to Equations (6) and (7). A series of reactions can happen with the generation of other active radicals at different pH values, such as •OH, O_2^-•, O_3^-•, HO_3•, O^-•, and HO_2•(Equations (8)–(13)). Therefore, the pH of the solution played an important role in diatrizoate degradation during ozonation (Gunten 2003; Virmani *et al.* 2020).

$$O_3 + OH^- \rightarrow HO_2^- + O_2, \quad k = 4.2 \times 10^2\ M^{-1}\ min^{-1} \tag{6}$$

$$O_3 + HO_2^- \rightarrow \bullet OH + O_2^-\bullet + O_2, \quad k = 1.68 \times 10^8\ M^{-1}\ min^{-1} \tag{7}$$

$$O_3 + O_2^-\bullet \rightarrow O_3^-\bullet + O_2, \quad k = 9.6 \times 10^{10}\ M^{-1}\ min^{-1} \tag{8}$$

$$\text{at pH} \leq 7.5: \quad O_3^-\bullet + H^+ \rightleftharpoons HO_3\bullet \tag{9}$$

$$HO_3\bullet \rightarrow \bullet OH + O_2 \tag{10}$$

$$\text{at pH} \geq 7.5: \quad O_3^-\bullet \rightleftharpoons O^-\bullet + O_2 \tag{11}$$

$$O^-\bullet + H_2O \rightarrow \bullet OH + OH^- \tag{12}$$

$$\bullet OH + O_3 \rightarrow HO_2\bullet + O_2 \tag{13}$$

3.4. Effects of bromide concentration on diatrizoate degradation during ozonation

Figure 5(a) and 5(b) displays the effect of bromide concentration on the degradation of diatrizoate. It is obvious that the rate of diatrizoate degradation significantly decreased as the bromide concentration increased in the solution. The reason is that Br^- can compete with diatrizoate for ozone consumption and cause a series of chain reactions (Equations (14)–(20); Song *et al.* 1996). Solutions containing Br^- are prone to form BrO_3^- during ozonation (Equation (18)), which has weak oxidizing power for organic compounds (Gunten 2003; Nie *et al.* 2013). Although in Equation (16), a strong oxidant HOBr and several bromine radicals can be formed, a part of them can further react with •OH to form BrO^- and BrO_2^- (Equations (17)–(20)), which are also weak oxidants (Huang *et al.* 2020b). Therefore, the presence of bromide exhibited an inhibitory effect on diatrizoate degradation during ozonation.

$$2Br^- + 2O_3 \rightarrow 2Br\bullet + 3O_2 \tag{14}$$

$$Br\bullet + Br^- \rightarrow Br_2^-\bullet, \quad k = 6 \times 10^{12}\ M^{-1}\ min^{-1} \tag{15}$$

$$Br_2^-\bullet + 3O_3 + H_2O \rightarrow 2HOBr + 4O_2 \tag{16}$$

$$HOBr + O_3 \rightarrow HBrO_2 + O_2 \tag{17}$$

$$HOBr + 2O_3 \rightarrow HBrO_3 + 2O_2 \tag{18}$$

$$Br_2^-\bullet + Br_2^-\bullet \rightarrow Br^- + Br_3^-\bullet, \quad k = 1.2 \times 10^{11}\ M^{-1}\ min^{-1} \tag{19}$$

$$Br_3^-\bullet + H_2O \rightarrow HOBr + H^+ + 2Br^- \tag{20}$$

3.5. Effect of solution pH on DBP formation of ozonated diatrizoate in the sequential chlorination

In previous literature, pH can have significant effects on DBP formation during disinfection (especially using chlorine). Therefore, DBP formation potential of diatrizoate after ozonation was also evaluated after 3 d chlorination to simulate water distribution in pipelines. As shown in Figure 6 in the absence of bromide, only trichloromethane ($CHCl_3$) was detected in the sequential chlorine process. Due to the strong oxidizing ability of ozone and •OH, diatrizoate can be decomposed into small molecular structures. ICMs can be degraded to a different extent during AOPs to form inorganic iodine and organic iodide at pH >7 (Ina *et al.* 2009), and ozone can oxidize iodide ions to HOI (Equations (21) and (22)) that can further react with O_3 to generate IO_2^- and IO_3^- (Equations (23) and (24)). Therefore, iodate will be the major product during

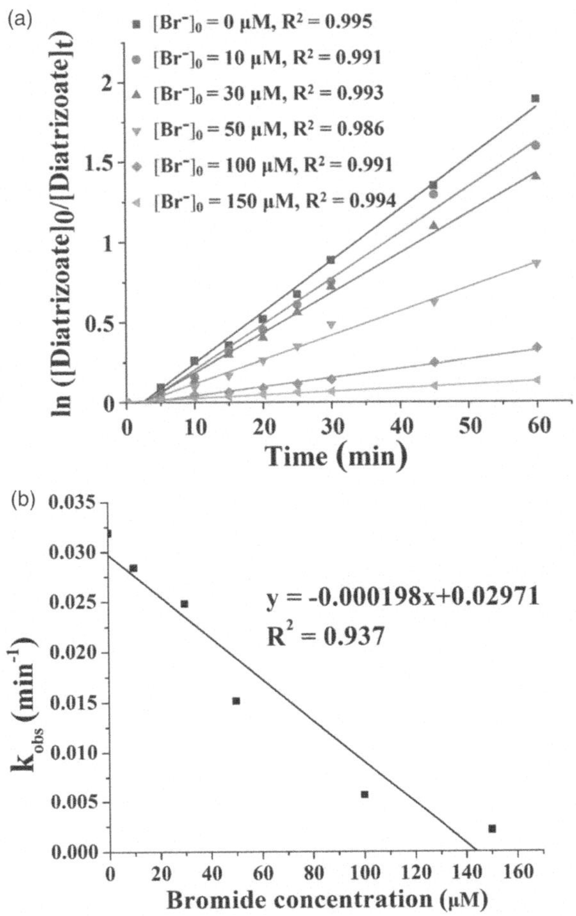

Figure 5 | (a) Pseudo-first-order kinetics plot of diatrizoate degradation at different bromine concentrations during ozonation at pH 7, [ozone] = 0.713 mg/L, [diatrizoate]0 = 20 μM, [phosphate buffer]0 = 10 mM, and temperature = 25 °C. (b) Correlation between k_{obs} and bromide concentration.

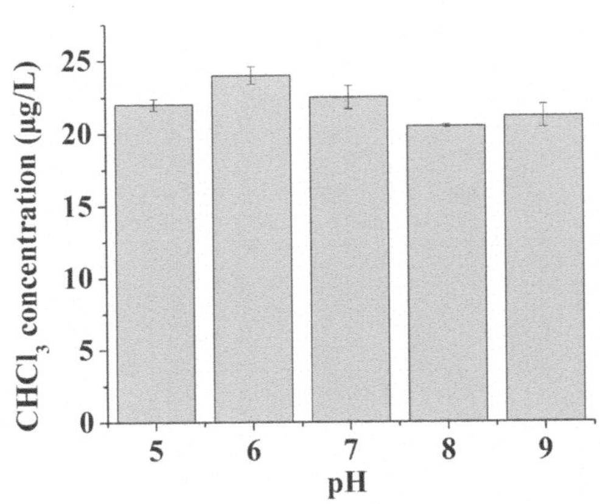

Figure 6 | DBP formation of ozone oxidized diatrizoate after 3 d chlorination at pH 5–9, [diatrizoate]0 = 20 μM, [ozone] = 0.713 mg/L, [HOCl] 0 = 100 μM, [phosphate buffer]0 = 10 mM, and temperature = 25 °C.

the ozone disinfection process without forming a considerable amount of I-DBPs (Von 2003).

$$I^- + O_3 \rightarrow OI^- + O_2, \quad k = 9.6 \times 10^7\ M^{-1}\ min^{-1} \tag{21}$$

$$H^+ + OI^- \rightarrow HOI, \quad k = 2.16 \times 10^6\ M^{-1}\ min^{-1} \tag{22}$$

$$OI^- + O_3 \rightarrow IO_2^- + O_2 \tag{23}$$

$$IO_2^- + O_3 \rightarrow IO_3^- + O_2 \tag{24}$$

3.6. Effects of bromide concentration on the formation of DBPs after post-chlorination

Figure 7 displays the effect of bromide concentration on the formation of DBPs after chlorination of ozone oxidized diatrizoate. It can be seen that six more kinds of DBPs besides $CHCl_3$ were detected, including $CHClBr_2$, $CHCl_2Br$, CHClBrI, $CHBr_2I$, C_2HBr_2N, and $CHBr_2Cl$, and the dominant one was CHClBrI. As the bromide concentration increased from 5 to 200 μM, C_2HCl_3, C_2HBrN, and $CHClBr_2$ formation remained relatively low and stable. However, the concentrations of $CHBr_3$ increased with increasing bromine concentration, while the concentrations of two detected I-DBPs, $CHBr_2I$, and CHClBrI exhibited an increasing and then decreasing trend. As the bromide concentration reached 100 μM, $CHBr_2I$ was detected. Accordingly, the presence of Br^- has a considerable impact on the production of DBPs. In order to analyze the contribution of bromine to the formation of THMs, the bromine incorporation factor (BIF) was calculated according to the following equation:

$$BIF = Br^{atom}_{number\ x}[Br - THMs]/[total\ THM\ species] \tag{25}$$

The range of BIF is 0–3 (Rathbun 1996). The larger the BIF value, the greater the contribution of bromine to the formation of THMs, indicating the greater proportion of Br-THMs in the total concentration of THMs. As the concentration of bromide increased from 10 to 200 μM, the BIF increased from 1.15 to 2.01, implying that Br-DBPs were preferentially produced, and chlorinated-DBPs were inhibited in the presence of bromide in the subsequent chlorination after diatrizoate ozonation (Ates *et al.* 2007). The presence of Br^- in the solution can react with HOCl to form HOBr (Cowman & Singer 1996; Brix *et al.* 2017), which is a strong oxidizing agent with oxidation power stronger than that of HOCl (Symons *et al.* 1993; Maryam *et al.* 2020) and can react with diatrizoate or its degradation intermediates at higher reaction rates (Chang *et al.* 2001), causing more formation of Br-DBPs (and even I-DBPs) with higher toxicity. Therefore, although ozone is efficient to decompose diatrizoate, it is essential to control the potential risks of forming toxic Br-DBPs and I-DBPs in the presence of bromide in the sequential chlorination.

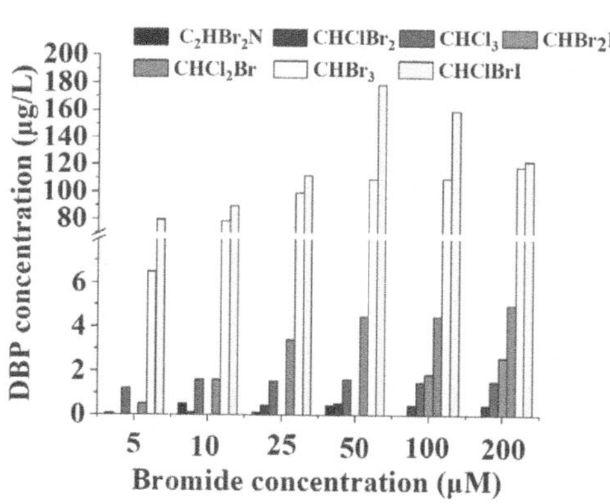

Figure 7 | DBP formation of oxidized diatrizoate after 3 d chlorination ozonation at different bromide concentrations, [diatrizoate]0 = 20 μM, [ozone] = 0.713 mg/L, [HOCl]0 = 100 μM, [phosphate buffer]0 = 10 mM, pH = 7, and temperature = 25 °C.

4. CONCLUSION

This study shows the successful application of ozone to completely decompose diatrizoate in alkaline conditions. Diatrizoate degradation during ozonation followed pseudo-first-order kinetics well. •OH played a major role in the diatrizoate degradation process, which was confirmed by the quenching experiments using TBA. At 25 °C, the degradation rate of diatrizoate by ozonation increased with increasing solution pH and ozone concentration. In the sequential chlorination simulating for water distribution in the pipelines, only one DBP, $CHCl_3$, was detected, and its formation was not pH-dependent. On the other hand, as Br^- concentration increased, the diatrizoate degradation efficiency decreased significantly due to the competition of ozone with diatrizoate, and six more kinds of DBPs were detected in the water after post-chlorination. Moreover, as Br^- concentration increased to 100 μM, Br-DBPs were preferentially produced, and $CHBr_2I$ was detected. Therefore, although ozone is efficient to decompose diatrizoate, it is essential to control the potential risks of forming toxic Br-DBPs and I-DBPs in the presence of bromide in the sequential chlorination.

ACKNOWLEDGEMENTS

This study was supported by the Natural Science Foundation of China (Grant No. 52170006), Shanghai Committee of Science and Technology, China (Grant No. 17DZ2282800), and the Ministry of Science and Technology, Taiwan (MOST-110-2221-E-992-025).

DATA AVAILABILITY STATEMENT

All relevant data are included in the paper or its Supplementary Information.

REFERENCES

Allard, S., Nottle, C. E., Chan, A., Joll, C. & Gunten, U. V. 2013 Ozonation of iodide-containing waters: selective oxidation of iodide to iodate with simultaneous minimization of bromate and I-THMs. *Water Res.* **47**, 1953–1960.

Ates, N., Yetis, U. & Kitis, M. 2007 Effects of bromide ion and natural organic matter fractions on the formation and speciation of chlorination by-products. *J. Environ. Eng.* **133**, 947–954.

Azerrad, S. P., Lutke Eversloh, C., Gilboa, M., Schulz, M., Ternes, T. & Dosoretz, C. G. 2016 Identification of transformation products during advanced oxidation of diatrizoate: effect of water matrix and oxidation process. *Water Res.* **103**, 424–434.

Brix, K., Hein, C., Sander, J. M. & Kautenburger, R. 2017 Simultaneous quantification of iodine and high valiant metals via ICP-MS under acidic conditions in complex matrices. *Talanta* **167**, 532–536.

Chang, E. E., Lin, Y. P. & Chiang, P. C. 2001 Effects of bromide on the formation of THMs and HAAs. *Chemosphere* **43**, 1029–1034.

Cowman, G. A. & Singer, P. C. 1996 Effect of bromide ion on haloacetic acid speciation resulting from chlorination and chloramination of aquatic humic substances. *Environ. Sci. Technol.* **30**, 16–24.

Duirk, S. E., Lindell, C., Cornelison, C. C., Kormos, J., Ternes, T. A., Attene-Ramos, M., Osiol, J., Wagner, E. D., Plewa, M. J. & Richardson, S. D. 2011 Formation of toxic iodinated disinfection by-products from compounds used in medical imaging. *Environ. Sci. Technol.* **45** (16), 6845–6854.

Gharekhanloo, F. & Torabian, S. 2012 Comparison of allergic adverse effects and contrast enhancement between iodixanol and iopromide. *Iran J. Radiol.* **9** (2), 63–66.

Giuseppe, M., Antonio, L., Huw, J. & Mike, F. 2007 By-products formation during degradation of isoproturon in aqueous solution. i: ozonation. *Water Res.* **7**, 1695–1704.

Gunten, U. 2003 Ozonation of drinking water: Part I. Oxidation kinetics and product formation. *Water Res.* **37** (7), 1443–1467.

Hennebel, T., De Corte, S., Vanhaecke, L., Vanherck, K., Forrez, I., De Gusseme, B., Verhagen, P., Verbeken, K., Van der Bruggen, B., Vankelecom, I., Boon, N. & Verstraete, W. 2010 Removal of diatrizoate with catalytically active membranes incorporating microbially produced palladium nanoparticles. *Water Res.* **44** (5), 1498–1506.

Hu, C. Y., Hou, Y. Z., Lin, Y. L., Li, A. P. & Deng, Y. G. 2019a Degradation kinetics of diatrizoate during UV photolysis and UV/chlorination. *Chem. Eng. J.* **360**, 1003–1010.

Hu, C. Y., Hou, Y. Z., Lin, Y. L., Deng, Y. G., Hua, S. J., Du, Y. F., Chen, C. W. & Wu, C. H. 2019b Kinetics and model development of iohexol degradation during UV/H_2O_2 and $UV/S_2O_8^{2-}$ oxidation. *Chemosphere* **229**, 602–610.

Hu, C. Y., Du, Y. F., Lin, Y. L., Hua, S. J., Hou, Y. Z., Deng, Y. G., Zhang, J. C., Ren, S. C., Dong, C. D., Chen, C. W. & Wu, C. H. 2020a Kinetics of iohexol degradation by ozonation and formation of DBPs during post-chlorination. *J. Water Process Eng.* **35**, 2214–7144.

Hu, C. Y., Hua, S. J., Lin, Y. L., Deng, Y. G., Hou, Y. Z., Du, Y. F., Dong, C. D., Chen, C. W. & Wu, C. H. 2020b Kinetics and formation of disinfection byproducts during iohexol chlor(am)ination. *Sep. Purif. Technol.* **243**, 1383–5866.

Huang, X. X., Quan, X. J., Cheng, W. & Li, J. 2020a Enhancement of ozone mass transfer by stainless steel wire mesh and its effect on hydroxyl radical generation. *Ozone Sci. Eng.* **42**, 347–356.

Huang, Y., Luo, M., Li, S., Xia, D., Tang, Z., Hu, S., Ye, S., Sun, M., He, C. & Shu, D. 2020b Efficient catalytic activity and bromate minimization over lattice oxygen-rich MnOOH nanorods in catalytic ozonation of bromide-containing organic pollutants: lattice oxygen-directed redox cycle and bromate reduction. *J. Hazard. Mater.* **410**, 124545.

Ina, K., Hervé, G., Cynthia, J. & Jean-Philippe, C. 2009 The formation of halogen-specific TOX from chlorination and chloramination of natural organic matter isolates. *Water Res.* **43**, 4177–4186.

Jeong, C. H., Machek, E. J., Shakeri, M., Duirk, S. E., Ternes, T. A., Richardson, S. D., Wagner, E. D. & Plewa, M. J. 2017 The impact of iodinated X-ray contrast agents on formation and toxicity of disinfection by-products in drinking water. *J. Environ. Sci.* **58**, 173–182.

Jiang, J., Zhang, X., Zhu, X. & Li, Y. 2017 Removal of intermediate aromatic halogenated DBPs by activated carbon adsorption: a new approach to controlling halogenated DBPs in chlorinated drinking water. *Environ. Sci. Technol.* **51** (6), 3435–3444.

Jiang, J., Han, J. & Zhang, X. 2020 Nonhalogenated aromatic DBPs in drinking water chlorination: a gap between NOM and halogenated aromatic DBPs. *Environ. Sci. Technol.* **54** (3), 1646–1656.

Kormos, J. L., Schulz, M. & Ternes, T. A. 2011 Occurrence of iodinated X-ray contrast media and their biotransformation products in the urban water cycle. *Environ. Sci. Technol.* **45** (20), 8723–8732.

Li, X. F. & Mitch, W. A. 2018 Drinking water disinfection byproducts (DBPs) and human health effects: multidisciplinary challenges and opportunities. *Environ. Sci. Technol.* **52** (4), 1681–1689.

Liu, Z., Lin, Y. L., Xu, B., Hu, C. Y., Wang, A. Q., Gao, Z. C., Xia, S. J. & Gao, N. Y. 2018 Formation of iodinated trihalomethanes during breakpoint chlorination of iodide-containing water. *J. Hazard. Mater.* **353**, 505–513.

Mao, Y. X., Dong, H. Y., Liu, S. G., Zhang, L. P. & Qiang, Z. M. 2020 Accelerated oxidation of iopamidol by ozone/peroxymonosulfate (O_3/PMS) process: kinetics, mechanism and simultaneous reduction of iodinated disinfection by-product formation potential. *Water Res.* **173**, 115615.

Maryam, K., Ben, C., Cordwell, S. J., David, I. P. & Michael, J. D. 2020 Characterization of disulfide (cystine) oxidation by HOCl in a model peptide: evidence for oxygen addition, disulfide bond cleavage and adduct formation with thiols. *Free Radic. Biol. Med.* **154**, 62–74.

Meng, L., Yang, S., Sun, C., He, H., Xian, Q., Li, S., Wang, G., Zhang, L. & Jiang, D. 2017 A novel method for photo-oxidative degradation of diatrizoate in water via electromagnetic induction electrodeless lamp. *J. Hazard. Mater.* **337**, 34–46.

Michael, J. P., Mark, G. M., Susan, D. R., Francesca, F. & Occurrence, M. B. 2008 Synthesis, and mammalian cell cytotoxicity and genotoxicity of haloacetamides: an emerging class of nitrogenous drinking water disinfection byproducts. *Environ. Sci. Technol.* **42**, 955–961.

Nie, Y., Hu, C., Yang, L. & Hu, J. 2013 Inhibition mechanism of formation over MnO_x/Al_2O_3 during the catalytic ozonation of 2,4-dichlorophenoxyacetic acid in water. *Sep. Purif. Technol.* **117**, 41–45.

Perez, S. & Barcelo, D. 2007 Fate and occurrence of X-ray contrast media in the environment. *Anal. Bioanal. Chem.* **387** (4), 1235–1246.

Polo, A. M. S., López-Peñalver, J. J., Sánchez-Polo, M., Rivera-Utrilla, J., Velo-Gala, I. & Salazar-Rábago, J. J. 2016 Oxidation of diatrizoate in aqueous phase by advanced oxidation processes based on solar radiation. *J. Photochem. Photobiol. A: Chem.* **319–320**, 87–95.

Ranjan, J., Mandal, T. & Mandal, D. D. 2021 Mechanistic insight for DBP induced growth inhibition in *Vigna radiata* via oxidative stress and DNA damage. *Chemosphere* **263**, 128062.

Rathbun, R. E. 1996 Bromine incorporation factors for trihalomethane formation for the Mississippi, Missouri, and Ohio rivers. *Sci. Total Environ.* **192**, 111–118.

Richardson, S. D., Plewa, M. J., Wagner, E. D., Schoeny, R. & Demarini, D. M. 2007 Occurrence, genotoxicity, and carcinogenicity of regulated and emerging disinfection by-products in drinking water: a review and roadmap for research. *Mutat. Res.* **636**, 178–242.

Sandra Perez, P. E., Celiz, M. D. & Aga, D. S. 2006 Structural characterization of metabolites of the X-ray contrast agent iopromide in activated sludge using ion trap mass spectrometry. *Anal. Chem.* **78**, 1866–1874.

Shang, W., Dong, Z., Li, M., Song, X., Zhang, M., Jiang, C. & Feiyun, S. 2019 Degradation of diatrizoate in water by Fe(II)-activated persulfate oxidation. *Chem. Eng. J.* **361**, 1333–1344.

Song, R., Donohoe, C., Minear, R., Westerhoff, P., Ozekin, K. & Amy, G. 1996 Empirical modeling of bromate formation during ozonation of bromide-containing waters. *Water Res.* **30**, 1161–1168.

Sugihara, M. N., Moeller, D., Paul, T. & Strathmann, T. J. 2013 TiO_2-photocatalyzed transformation of the recalcitrant X-ray contrast agent diatrizoate. *Appl. Catalysis B: Environ.* **129**, 114–122.

Symons, J. M., Krasner, S. W., Simms, L. A. & Sclimenti, M. 1993 Measurement of THM and precursor concentrations revisited: the effect of bromide ion. *J. AWWA* **85**, 51–62.

Ternes, T. A., Stüber, J., Herrmann, N., McDowell, D., Ried, A., Kampmann, M. & Teiser, B. 2003 Ozonation: a tool for removal of pharmaceuticals, contrast media and musk fragrances from wastewater. *Water Res.* **37** (8), 1976–1982.

Velo-Gala, I., López-Peñalver, J. J., Sánchez-Polo, M. & Rivera-Utrilla, J. 2014 Comparative study of oxidative degradation of sodium diatrizoate in aqueous solution by H_2O_2/Fe^{2+}, H_2O_2/Fe^{3+}, Fe(VI) and UV, H_2O_2/UV, $K_2S_2O_8$/UV. *Chem. Eng. J.* **241**, 504–512.

Virmani, A., Walavalkar, M. P., Sharma, A., Sengupta, S., Saha, A. & Kumar, A. 2020 Kinetic studies of the gas phase reaction of 1,2-propylene oxide with the OH radical over a temperature range of 261–335 K. *Atmos. Environ.* **237**, 117709.

Von, G. U. 2003 Ozonation of drinking water: part II. Disinfection and by-product formation in presence of bromide, iodide or chlorine. *Water Res.* **37**, 1469–1487.

Wang, Z., Lin, Y.-L., Xu, B., Xia, S.-J., Zhang, T.-Y. & Gao, N.-Y. 2016 Degradation of iohexol by UV/chlorine process and formation of iodinated trihalomethanes during post-chlorination. *Chem. Eng. J.* **283**, 1090–1096.

Xia, Y., Lin, Y. L., Xu, B., Hu, C. Y., Gao, Z. C., Chu, W. H. & Gao, N. Y. 2017 Iodinated trihalomethane formation during chloramination of iodate-containing waters in the presence of zero valent iron. *Water Res.* **124**, 219–226.
Yolanta, G., Tan, J., Allard, S., Heitz, A. & Bowman, M. 2014 Impact of bromide and iodide during drinking water disinfection and potential treatment processes for their removal or mitigation. *Water* **41**, 38–43.
Zhang, M., Wang, X., Hao, H., Wang, H., Duan, L. & Li, Y. 2019 Formation of disinfection byproducts as affected by biochar during water treatment. *Chemosphere* **233**, 190–197.

First received 15 June 2021; accepted in revised form 13 September 2021. Available online 30 September 2021

Water Recycling and Low-Carbon Development doi: 10.2166/wrd.2021.049

The *in situ* remediation of aquaculture water and sediment by commercial probiotics immobilized on different carriers

Mengmeng Yi[a], Chun Wang[b], He Wang[a], Xi Zhu[a], Zhigang Liu[a], Fengying Gao[a], Xiaoli Ke[a], Jianmeng Cao[a], Miao Wang[a], Ying Liu[c] and Maixin Lu[a,*]

[a] Key Laboratory of Tropical and Subtropical Fishery Resource Application and Cultivation, Ministry of Agriculture, Guangdong Provincial Key Laboratory of Aquatic Animal Immune Technology, Pearl River Fisheries Research Institute, Chinese Academy of Fishery Science, Guangzhou 510380, China
[b] State Environmental Protection Key Laboratory of Food Chain Pollution Control, Beijing Technology and Business University, Beijing 100048, China
[c] Key Laboratory of Environment Controlled Aquaculture, Ministry of Education, Dalian Ocean University, Dalian 116023, China
*Corresponding author. E-mail: mx-lu@163.com

MY, 0000-0001-9220-6943; ZL, 0000-0002-1815-1101; JC, 0000-0001-5035-3814

ABSTRACT

In the present study, we investigated the effect of probiotics immobilized by oyster shells (Os), vesuvianite (Ve) and walnut shells (Ws) on the remediation of aquaculture water and sediment by analyzing the variation of ammonia-nitrogen (NH_4–N), nitrate-nitrogen (NO_3–N), nitrite-nitrogen (NO_2–N), total nitrogen (TN), total phosphorus (TP) and chemical oxygen demand (COD_{Cr}), as well as the microbiota of the water and sediment. The positive or negative effects of the treatment groups on the water quality parameters were both observed. Compared with their effects on water quality parameters, the treatment groups had better effects on sediment parameters. Group Ve had the best remediation effect of NH_4–N and NO_3–N in the sediment (decreased by 5.22 and 1.66 times, respectively). Group Os showed a lower relative concentration of TN and COD_{Cr} (decreased by 3.77 and 0.95 times, respectively). The high-throughput sequencing results revealed that the immobilized probiotics increased the relative abundances of functional bacteria in the treatment groups at the phylum and genus level. The above results showed that probiotics immobilized by oyster shells, vesuvianite and walnut shells positively affected the aquaculture environment's remediation, especially the sediment.

Key words: aquaculture water, immobilized probiotic, *in situ* remediation, microbial community

HIGHLIGHTS

- The effects of probiotics immobilized by oyster shells, vesuvianite and walnut shells on the remediation of aquaculture water and sediment were researched.
- The effect of immobilized probiotics on nutrient content and microbiota in the water and sediment was tested.
- Probiotics immobilized by oyster shells, vesuvianite and walnut shells affect the remediation of aquaculture water and sediment.

INTRODUCTION

Globally, 48 million tons of aquatic feed was produced in 2015, and the output is expected to reach 71 million tons by 2020 (FAO 2012, 2018), while approximately 50–80% of the nitrogen in feed is released into breeding ponds (Schneider *et al.* 2005). With the increases in aquaculture production and aquaculture area, the amount of eutrophic aquaculture wastewater to be treated dramatically increases. Physical, chemical and biological methods for remediation of nutrient-rich water and sediment have been applied and investigated for many years (Wang *et al.* 2017). As an eco-friendly and cost-effective biological method, probiotics for environmental remediation in aquaculture systems have gained more attention (Zhu *et al.* 2019). While due to the resulting low growth rate and unstable effect of probiotics, one of the main challenges for carrying out environmental remediation with probiotics is to ensure their retention inside aquaculture systems. Fortunately, this problem can be solved by utilizing immobilization. The immobilized biomass could increase the biodegradation rate, enhanced the control of the bioprocess, improved biocatalyst stability and increased the tolerance of harsh environmental conditions (Covarrubias *et al.* 2012). For these reasons, immobilized bacteria have received more attention in aquaculture environmental remediation to minimize the risk of washed-out bacterial cells from the systems and provide a stable effect. Therefore, selecting and producing the ideal carriers are essential for the immobilization of probiotics.

The ideal immobilized carriers should be chemically inert, cost-efficient, readily available and renewable and should have a favorable adsorptive capability. Oyster shells, composed of biogenic carbonate, are an efficient adsorbent for removing some pollutants. Many studies have used oyster shells in wastewater remediation (Shih & Chang 2015; Yen & Chou 2016). Oyster shells showed good performance in removing phosphate (Oladoja *et al.* 2015), hydrogen sulfide (Asaoka *et al.* 2009) and dissolved contaminant cations (Wu *et al.* 2014) from wastewater benefit from its porous structure and high calcium oxide content after being thermally treated. The rough surface and sufficient alkalinity of the oyster shell is conducive to microbe growth. These characteristics make it a promising media with higher NH_3-N removal efficiency in biological aerated filters (Liu *et al.* 2010). Oyster cultivation on the coast of China exceeded 5.14 million tons in 2018 (Ministry of Agriculture and Rural Affairs of the People's Republic of China 2019). However, the discarded oyster shells caused environmental and health problems in coastal regions (Khan *et al.* 2018). Therefore, the reuse of waste oyster shells as an immobilization carrier of probiotics is eco-friendly and economically viable. Similarly, approximately 67% of the walnut weight is accounted by shell, and walnut shells are a low-cost, abundant agricultural by-product (Pirayesh *et al.* 2012). Usually, walnut shells, a lignocellulosic material with good chemical stability and high mechanical strength, are discarded or burned. Recently, walnut shells with a high fixed carbon content are derived biochar which possesses high basicity, large pore volume and surface area (Qiu *et al.* 2018). Modified walnut shells biochar have been used as a catalyst to remove organic sulfur and arsenic (Song *et al.* 2017). Walnut shells have been successfully used in wastewater and sewage sludge remediation as well (Segovia-Sandoval *et al.* 2018; Wójcik 2020). The porosity of walnut shells forms the oxic and anoxic zones and provided enough space for bacteria growth to generate high cell density (Wójcik 2020). These characteristics make it possible for walnut shells to be applied as immobilization carriers. The vesuvianite is another promising immobilization carrier. Vesuvianite shows a superior capability to remove pollutants, such as nitrogen, phosphate organic matter and CH_4 in the water and sewage sludge, which might be attributed to their high total pore volume, high specific area and low density (Li *et al.* 2009; Jiang *et al.* 2019; Kang *et al.* 2019).

The reuse of oyster shells and walnut shells as immobilized carriers for probiotics might be a good way to add value to this aquaculture and agricultural waste and increase the aquaculture water and sediment remediation efficiency of commercial probiotics. Furthermore, there has been no research so far taking oyster shells, vesuvianite and walnut shells as immobilized carriers for probiotics. Therefore, oyster shells, vesuvianite and walnut shells were selected as immobilized carriers for probiotics in this research. Our research aimed to investigate the influence of probiotics immobilized by oyster shells, vesuvianite and walnut shells on the remediation of aquaculture water and sediment. We tested the effect of immobilized probiotics on nutrient content in the forms of ammonia-nitrogen (NH_4–N), nitrate-nitrogen (NO_3–N), nitrite-nitrogen (NO_2–N), total nitrogen (TN), total phosphorus (TP) and chemical oxygen demand (COD_{Cr}) in the water and sediment. To further visualize the probiotics immobilized by carriers and carriers' surface characterization, the scanning electron microscope (SEM) was applied. High-throughput sequencing was used to reveal the influence of immobilized probiotics on the microflora of water and sediment. The results of this research will verify the feasibility of further research concerning the improvement of oyster shells, vesuvianite and walnut shells as immobilized carriers for probiotics.

METHOD

Commercial probiotic immobilization and application

Four treatment groups (the control group, C; probiotics immobilized by oyster shells, Os; probiotics immobilized by vesuvianite, Ve; and probiotics immobilized by walnut shells, Ws) were studied (in triplicate per treatment) with 12 freshwater ponds. Sediment (10 cm) from a genetic improvement of farmed tilapia (*Oreochromis niloticus*, GIFT strain) pond covered the bottoms of the cement ponds (1.60 m length, 2.0 m width and 0.75 m depth). A total of 2.4 kg of the carrier, oyster shells, vesuvianite and walnut shells was loaded with 3.5 L of commercial probiotics for 6 h at room temperature. After immobilization, we added the carrier-immobilized probiotics to cement ponds. Furthermore, we added equal amounts of probiotics without a carrier to the control group. Oxygen was provided by a bottom microporous aeration system (approximately 8 cm above the top sediment layer). Temperature, pH and dissolved oxygen (DO) were measured daily by a YSI 6600 data sonde (YSI, USA).

SEM observation

A single colony of *Bacillus cereus* (NY5 strain, ingredient of the commercial probiotics) was inoculated with an LB culture medium at 200 rpm, 30 °C for 8 h. 5 mL 10^8 CFU/mL of NY5 were immobilized by 1 g oyster shells, vesuvianite and walnut shells, 30 °C for 12 h, respectively. Subsequently, 30, 50, 70, 80, 90, 95 and 100% ethanol were used to successive dehydration of immobilized NY5 fixed by glutaraldehyde (2.5%). The immobilized NY5 samples were coated with gold and then measured by SEM (S4800, Hitachi).

Water and sediment analyses

Water and sediment samples were collected from four locations in each tank at 0 and 12 weeks, as shown in Figure 1. Then, the samples in each tank were mixed and stored in polyethene plastic bottles. The samples were kept under low-temperature conditions, transported to the laboratory and stored in a 4 °C fridge for analysis within 24 h. The concentrations of NH_4–N, NO_3–N, NO_2–N, TN, TP and COD_{Cr} of the water samples were determined following standard procedures of Chinese Environmental Quality Standards (Nessler's reagent spectrophotometric method (HJ 535-2009), phenol disulfonic acid spectrophotometric method (GB 748-87), *N*-(1-naphthyl)-ethylenediamine dihydrochloride spectrophotometric method (GB 7493-87), alkaline potassium persulfate digestion spectrophotometric method (HJ 636-2012), ammonium molybdate spectrophotometric method (GB11893-89) and potassium dichromate method (GB 11914-89), respectively). Ten grams of sediment sample was added to 100 mL of KCl (1 mol/L) and shaken at 120 rpm for 1 h at 20 °C. Then, the samples were centrifuged at 3,000 rpm for 10 min to separate the supernatant from the solid. The supernatant was used to determine the NH_4^+-N, NO_3^--N, NO_2^--N, TN, TP and COD_{Cr} using the same methods for the water samples and KCl (1 mol/L) was used as a blank control.

The relative concentrations of water and sediment parameters (Equation (1)) were calculated for each treatment using the following equations:

$$C = C_{12w} - C_0 \tag{1}$$

where C_0 is the initial concentration (0 week) of the water and sediment parameters and C_{12w} is the concentration of the water and sediment parameters in the 12th week.

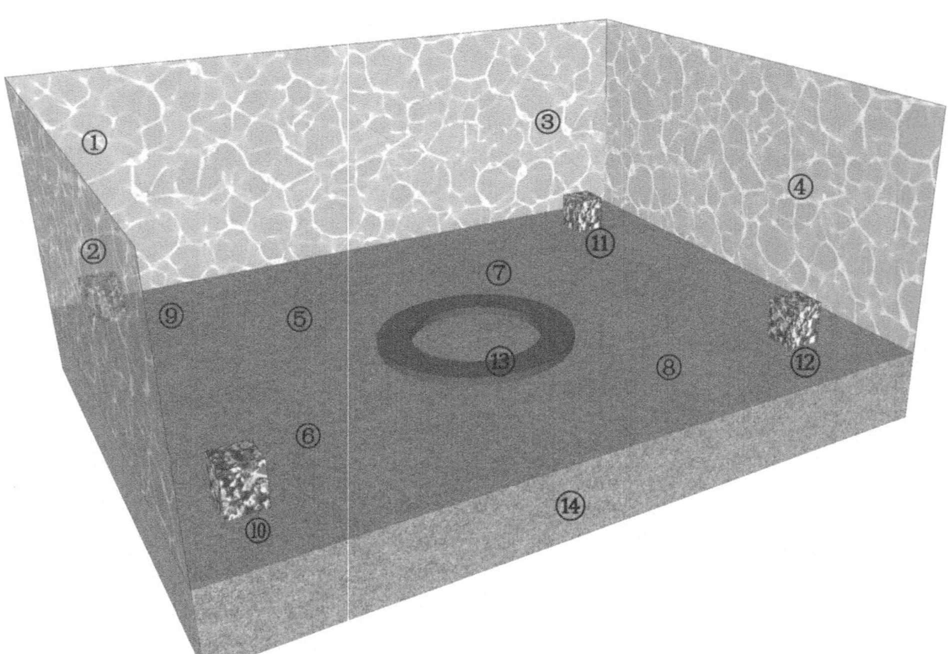

Figure 1 | The pond diagram, where ①, ②, ③, ④ were the water sampling location (20 cm underwater surface), ⑤, ⑥, ⑦, ⑧ were the sediment sampling location, ⑨, ⑩, ⑪, ⑫ were the carrier-immobilized probiotics, ⑬ was bottom microporous aeration system, and ⑭ was 10 cm sediment.

Water and sediment microbial DNA extraction, sequencing and analysis

Water samples (100 mL) were passed through a 0.22 μm polycarbonate membrane (GTTP04700, Merck Millipore, USA). Genomic DNA was extracted from filtrate and sediment samples by a DNeasy PowerSoil Kit (Qiagen, Carlsbad, CA). DNA quality was confirmed by 0.8% agarose gel electrophoresis and a spectrophotometer (NanoDrop Technologies Inc., Wilmington, DE, USA). To investigate the bacterial community structure of the environmental samples, DNA was amplified according to a protocol described previously (Nicholaus *et al.* 2019). The V4 region of the 16S rRNA gene was amplified with the 515f/806r primer set. The reverse primer contains a 6-bp error-correcting barcode unique to each sample. The amplified products of each sample were mixed, purified, equilibrated and sequenced on the Illumina MiSeq high-throughput sequencing platform.

To investigate the bacterial community structure of water and sediment samples, paired sequence reads from the original DNA fragments were merged using FLASH. Sequence reads from the same sample were given similar barcodes. Microbial phylotypes were identified with Quantitative Insights into Microbial Ecology (QIIME) software and the UPARSE pipeline (Gregory Caporaso *et al.* 2010). The number of OTUs in each sample was used to represent species richness.

Statistical analysis

The data were analyzed by IBM SPSS 22.0. The results for all groups were compared using a one-way analysis of variance (one-way ANOVA). In all cases, the significance level of differences was set at $P < 0.05$.

RESULTS AND DISCUSSION

The immobilization efficiency of different carriers

Probiotics with good denitrification potential have been widely used in aquaculture remediation (Nimrat *et al.* 2012; Christianson *et al.* 2016). More microbes with good denitrification potential have been isolated and investigated (Kewcharoen & Srisapoome 2019; Zhu *et al.* 2019). Comparing with free-living microorganisms, immobilization protected microorganisms from reducing the impact of the environment on them and enhanced their activity and stability (Covarrubias *et al.* 2012). In the current study, oyster shells, vesuvianite and walnut shells were used as probiotics immobilized. To further visualize the probiotics immobilized by carriers, the SEM observation was carried out. SEM images of carrier structure and immobilized NY5 by oyster shells, vesuvianite and walnut shells were observed (Figure 2). SEM images showed that the surface of oyster shells (Figure 2(a)) and walnut shells (Figure 2(c)) was relatively smooth, but the smooth surface does not affect their immobilizing the bacterial. The perfect immobilization ability of oyster shells may be attributed to the electrostatic interactions and higher the external surface area (Hrenovic *et al.* 2009). The previous research also demonstrated walnut shells were an efficient carrier for bacterial immobilization (Zare *et al.* 2012; Hasanzadeh *et al.* 2020). Inversely, although the surface of vesuvianite is full of macropores, the probiotics were not immobilized (Figure 2(b)). The remediation effect of group Ve was due to the porosity and higher the external surface area. These characters of vesuvianite are beneficial to the absorption of contaminants (Li *et al.* 2009; Jiang *et al.* 2019). According to the good immobilization performance of oyster shells and walnut shells, and the large specific surface area of vesuvianite, probiotics immobilized by oyster shells, vesuvianite and walnut shells were applied to aquaculture water and sediment remediation in the follow-up experiment.

Figure 2 | SEM images of carrier and immobilized NY5. (a) NY5 immobilized by oyster shells. (b) NY5 immobilized by vesuvianite. (c) NY5 immobilized by walnut shells. Scale bar: 5 μm.

The aquaculture water and sediment remediation efficiency of commercial probiotics immobilized by different carriers

Reducing the nutrients entering aquaculture water is urgent for the sustainable development of pond aquaculture. Bioremediation has been widely employed to purify nutrient-rich aquaculture water due to its lower cost than physicochemical methods (Xia *et al.* 2020). In the current study, we investigated water and sediment quality improvement via probiotics immobilized by oyster shells, vesuvianite and walnut shells. Figure 3 shows the relative concentration of NH_4–N in the water and sediment. After the 12 weeks of treatment, the relative NH_4–N concentration in Ws was significantly higher than in the C and Os groups ($P < 0.05$). While in the sediment, the relative NH_4–N concentration in Ve remained at a lower level than those in the other groups (decreased approximately 5.22 times compared with that in group C, $P < 0.05$), and that in Ws was only slightly lower than that in the C (decreased approximately 2.72 times, $P < 0.05$). A reverse NH_4–N treatment effect of the same group for water and sediment was observed, such as Group C and Os had a lower relative concentration of NH_4–N in the water, but the relative concentration of NH_4–N in the sediment of the two groups were higher compared with groups Ve and Ws.

The relative concentrations of NO_3–N in the water and sediment are shown in Figure 4. The relative concentration of NO_3–N in the C was lower than that in the other groups ($P < 0.05$), but there were no significant differences between other groups ($P > 0.05$). In the sediment, the relative NO_3–N concentration in Ws was significantly higher than those in the other groups. On the other hand, Ve was significantly lower than that in group C (decreased approximately 1.66 times, $P < 0.05$). A reverse NO_3–N treatment effect of the same group for water and sediment was also observed. For example, group Ve had the highest relative concentration of NO_3–N in the water, but the relative concentration of NO_3–N in the sediment was the lowest compared with other groups.

As shown in Figure 5, the relative NO_2–N concentrations of the four groups (C, Os, Ve and Ws) had significant differences. In water, the relative NO_2–N concentrations in Ws and Ve were lower than those in C (decreased approximately 0.92 and 0.86 times, respectively) and Os and were significantly different from that in C ($P < 0.05$). But there were no statistical significances between the four groups in sediment ($P > 0.05$).

The relative TN concentrations in the four groups (C, Os, Ve and Ws) after the 12 weeks of treatment are shown in Figure 6. At the end of the experiment, the relative TN concentration of group C in water was significantly lower than those in Os, Ve

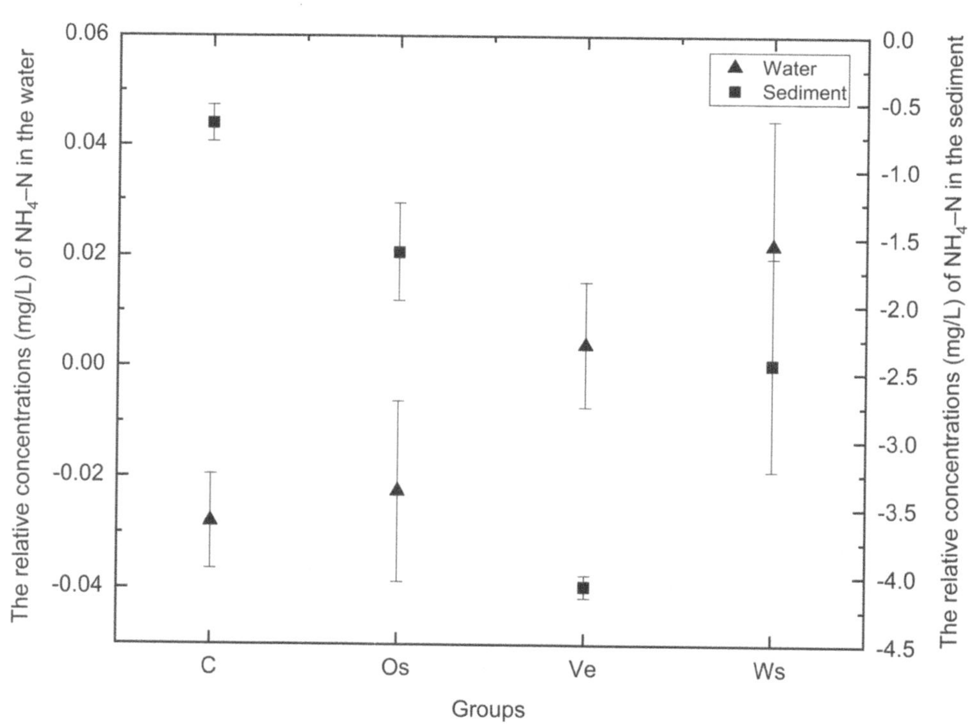

Figure 3 | The relative concentrations (mg/L) of NH_4–N in the water and sediment treated by commercial probiotics immobilized on different carriers.

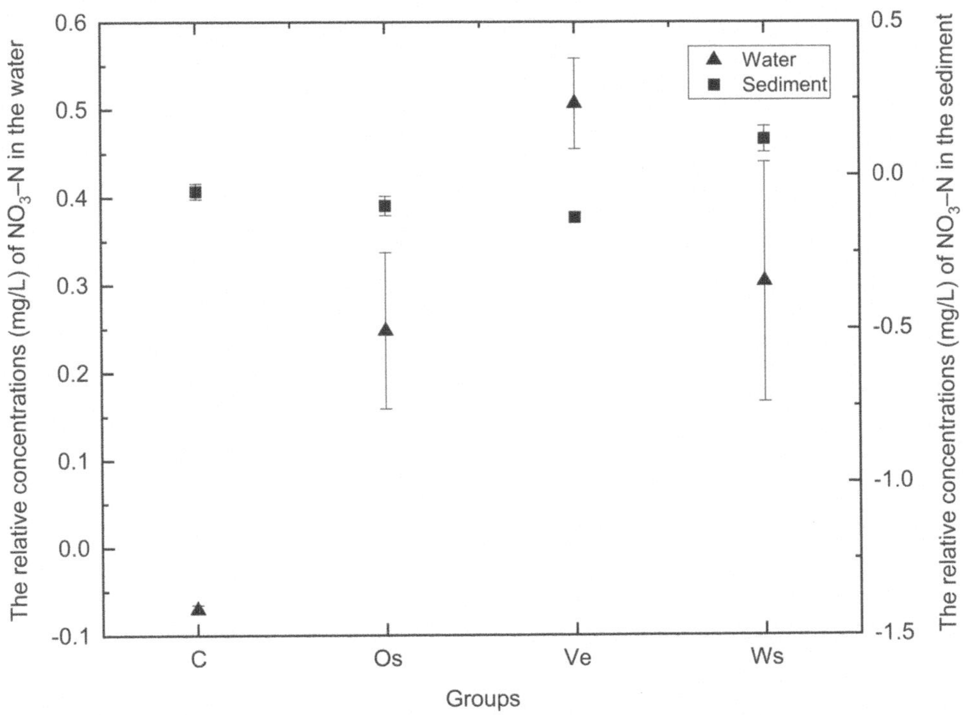

Figure 4 | The relative concentrations (mg/L) of NO_3–N in the water and sediment treated by commercial probiotics immobilized on different carriers.

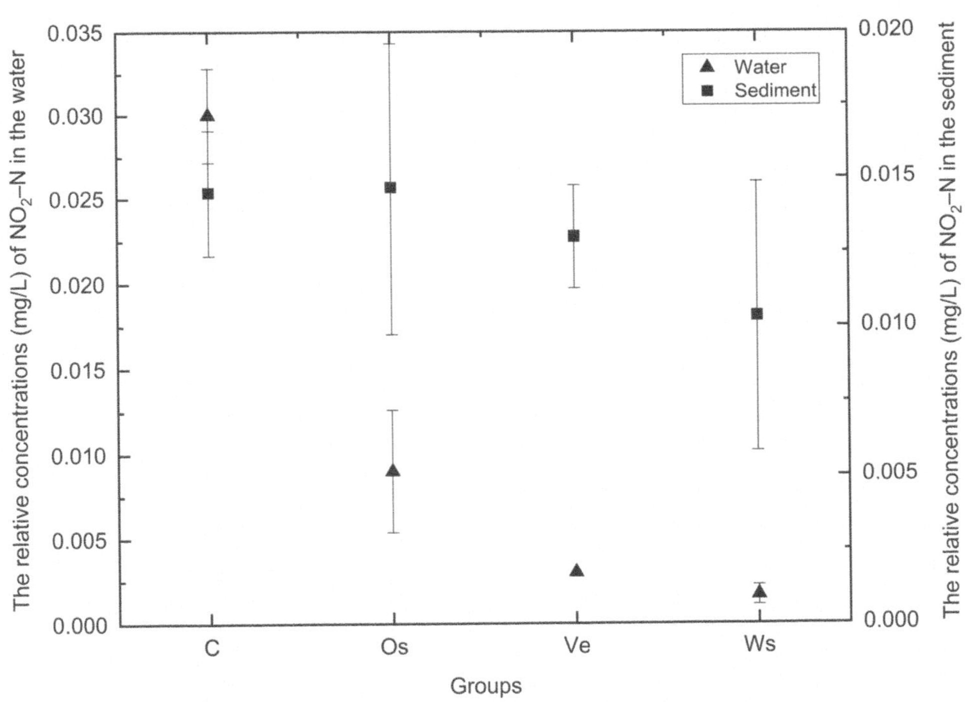

Figure 5 | The relative concentrations (mg/L) of NO_2–N in the water and sediment treated by commercial probiotics immobilized on different carriers.

and Ws ($P < 0.05$). In sediment, the concentration of group Os was significantly different from those in the other three groups (decreased approximately 3.77 times compared with that in group C, $P < 0.05$). A reverse TN treatment effect of the same group for water and sediment was also observed, such as Group Os had the highest relative concentration of TN in the

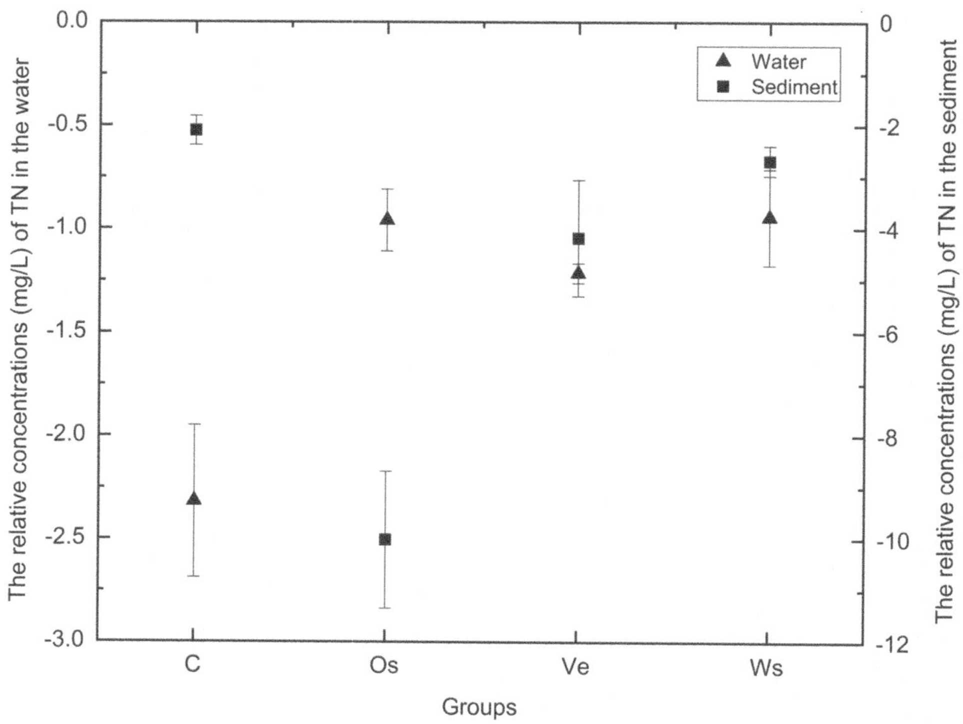

Figure 6 | The relative concentrations (mg/L) of TN in the water and sediment treated by commercial probiotics immobilized on different carriers.

water. Still, the relative TN concentration of group Os in the sediment was the lowest compared with other groups. Group C had the lowest relative concentration of TN in the water, but the relative TN concentration of group C in the sediment was the highest.

The relative concentrations of TP in the water and sediment are shown in Figure 7. At the end of the experiment, the relative concentrations in the C and Ws were significantly lower than those in Os and Ve ($P < 0.05$). In addition, the relative TP concentration of group C in sediment was lower than those in the other three groups and significantly lower than that of group Os and Ws ($P < 0.05$).

The relative concentrations of COD_{Cr} in the water and sediment are shown in Figure 8. There were no significant differences were observed between the four groups for COD_{Cr} treatment. However, in sediment, the highest relative concentration of COD_{Cr} was detected in the C, where it was significantly higher than in the other three groups (increased approximately 0.95 times compared with those in groups Os and Ve, $P < 0.05$).

Most of the residual feed, animal remains and excrement produced during aquaculture are deposited on the pond bottom and form parts of the sediment. And approximately 80–90% of the N and P input during aquaculture are deposited in the sediment (Schroeder 1987). The release of these nutrients into the water strongly influences the environmental quality of the aquaculture pond. Several strategies, such as bioremediation and physical treatment, have been used in the remediation of sediment. It was revealed that probiotic bacteria could enhance water quality (Nimrat *et al.* 2012). The addition of *Bacillus* directly to rearing water or biofloc could maintain the physicochemical parameters of water within the tolerable ranges for shrimp culture or elicit faster development of the biofloc and enhance water quality (Zokaeifar *et al.* 2014; Dash *et al.* 2018). Immobilized microbes were found to be more efficient than free cells (Fareed *et al.* 2019). Sodium carboxymethylcellulose anammox-immobilized sludge showed higher removal ratios of ammonium and nitrite (Zhu *et al.* 2009). Bacteria immobilized on wood chips, and maize straw exhibited more efficient biodegradation of oil in a marine environment than free bacteria (Xue *et al.* 2017). In our study, the relative concentrations of NH_4–N, NO_3–N, NO_2–N, TN and COD_{Cr} in the sediment were lower than those in water, indicating that the probiotic immobilized by vesuvianite, walnut shells and oyster shells decomposed the organic matter in the sediment. But part of the free N produced by the decomposition of the organic matter in the sediment was released into the aquaculture water. Accordingly compared with water quality parameters, a better effect on sediment parameters by immobilized probiotics was observed. Increased N content in the water of the immobilized

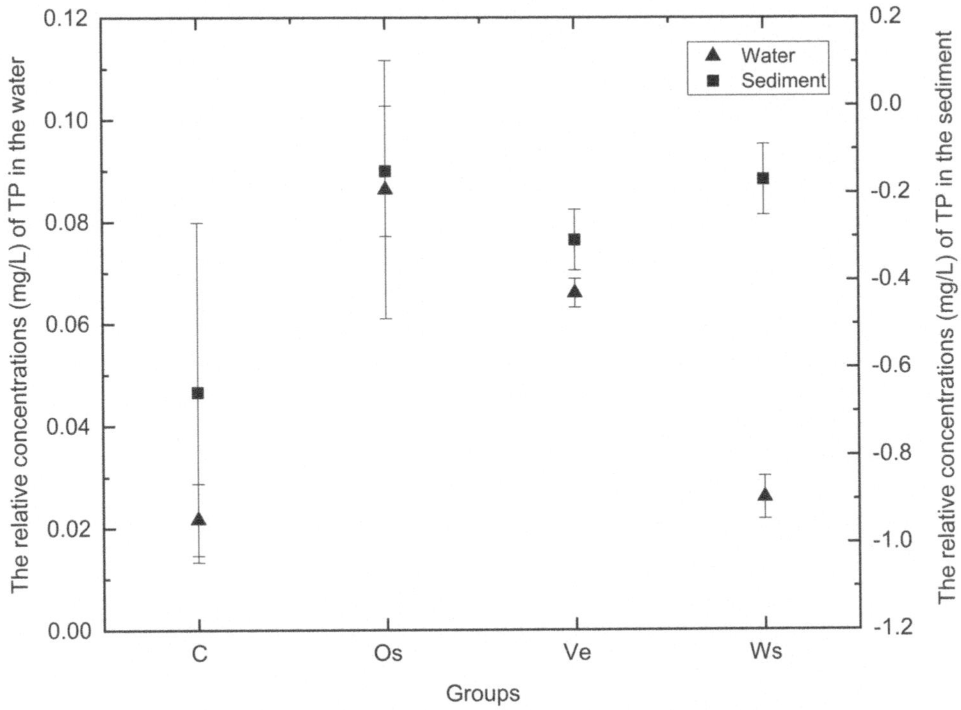

Figure 7 | The relative concentrations (mg/L) of TP in the water and sediment treated by commercial probiotics immobilized on different carriers.

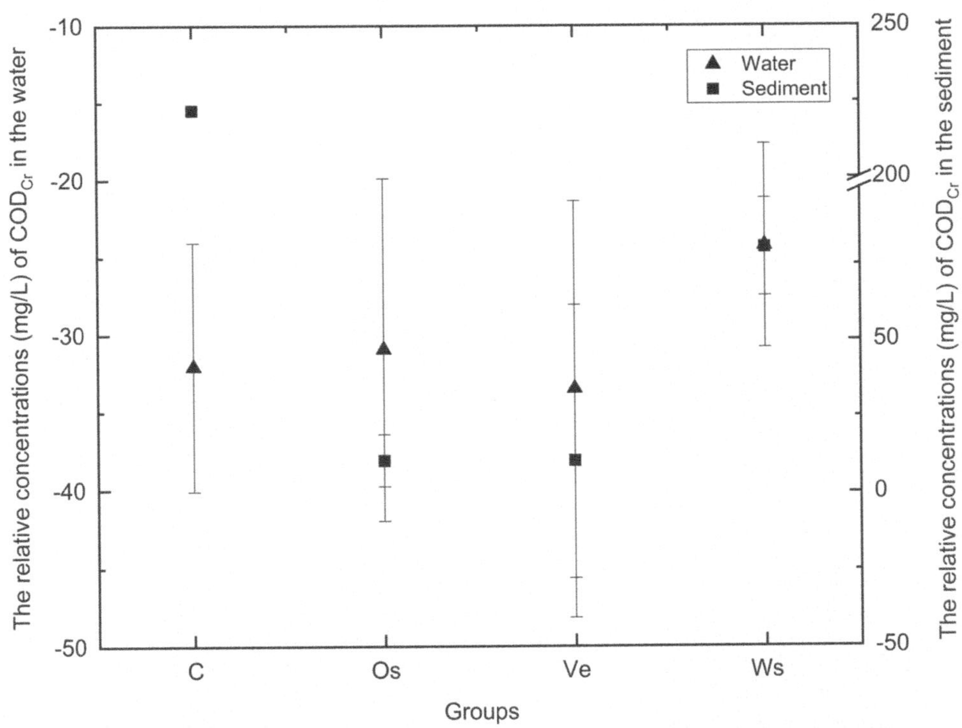

Figure 8 | The relative concentrations (mg/L) of COD_{Cr} in the water and sediment treated by commercial probiotics immobilized on different carriers.

probiotic groups was observed in the study. This is consistent with the finding of previous studies showing that organic nitrogen in the sediment is degraded during the initial days of investigation and the increase in nutrients in the overlying water was mainly due to ammonification by microbes (Fu *et al.* 2020). Similar results were found in which the release rate of nutrient fluxes from sediment was significantly increased due to bioturbation activities, in turn remediating the aquaculture effluents of blood clam (Nicholaus *et al.* 2019).

The remediation effects of immobilized probiotics were attributed to the activities of probiotics as well as the immobilization carriers. The results of SEM indicated that probiotics were not immobilized by vesuvianite, and group Ve showed lower relative concentrations of NH_4–N, TN and COD_{Cr} in sediment comparing with group C due to the absorption of vesuvianite. One study showed that the addition of vesuvianite accelerated organic matter degradation and decreased cumulative CH_4 emissions due to their special porosity microstructures (Jiang *et al.* 2019). Modified vesuvianite exhibited a high adsorption capacity for phosphate (Li *et al.* 2009). Other studies revealed that walnut shells are an efficient and cost-effective adsorbent for Cr (Banerjee *et al.* 2018), Cd (Qiu *et al.* 2018) and Zn (Segovia-Sandoval *et al.* 2018) removal, and the addition of ground walnut shells to sewage sludge reduce the total bacterial count and improved sewage sludge dewaterability (Wójcik 2020). Researches have also shown that oyster shells have a good water purification efficiency (Shih & Chang 2015; Jung *et al.* 2016). Therefore, the remediation efficiency of immobilized probiotics is a combined effect of probiotics and immobilization carriers attributed to their high mechanical strength and stability.

The effect of commercial probiotics immobilized by different carriers on the microbiota of the aquaculture environment

As the remediation effects of immobilized probiotics on sediment parameters were confirmed, high-throughput sequencing was used to investigate the structure and diversity of the bacterial community in water and sediment treated with probiotics immobilized by oyster shells, vesuvianite and walnut shells. The relative abundances of phyla detected with bacterial 16S rRNA sequences in the water and sediment samples in the different groups are shown in Figure 9. It is suggested that the presence and the activities exerted by immobilized probiotics are the reasons for the variation in microbial community composition. These results are supported by the finding of previous studies that reported considerable bacterial community change in water and sediment in response to probiotics (Nimrat *et al.* 2012; Chang *et al.* 2019; Nicholaus *et al.* 2019). Firmicutes was the most abundant phylum in the water of the C, with a relative abundance of 32.96% (Figure 9(a)). The

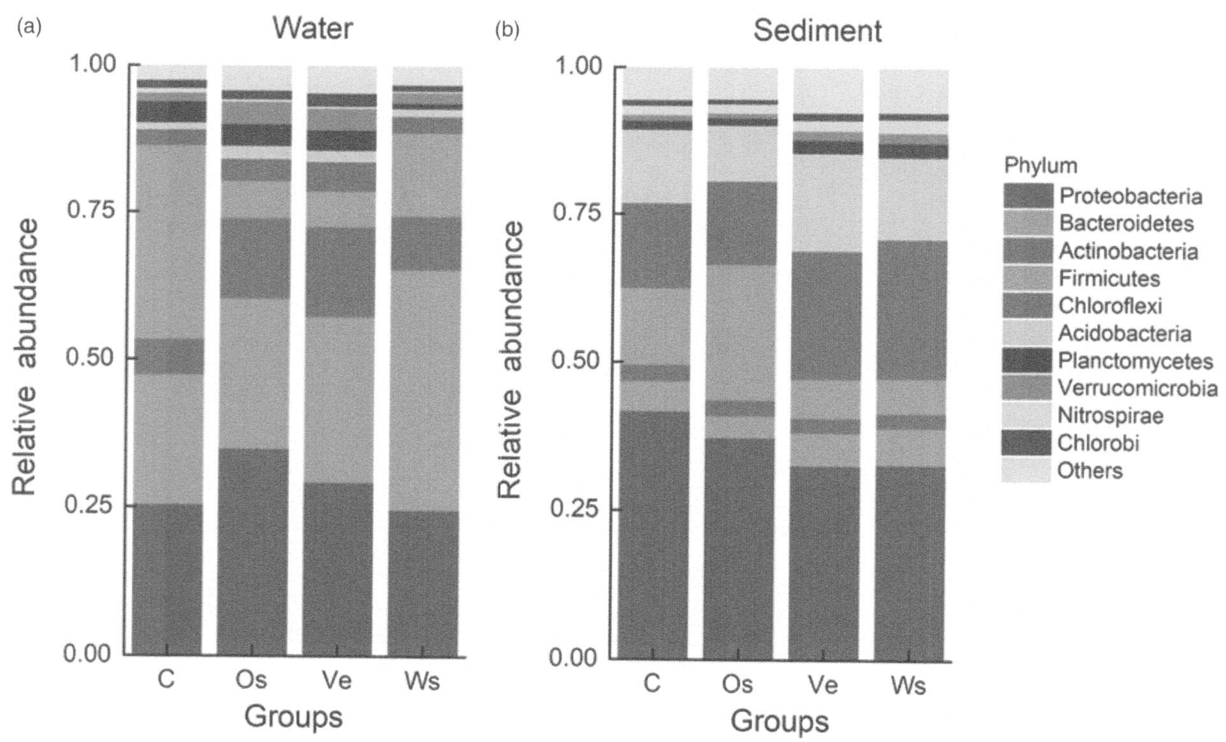

Figure 9 | The relative abundances of microbial community members (phylum level) of each group in the water (a) and sediment (b). 'Others' represents the phyla with a relative abundance lower than 0.01.

relative abundances of Proteobacteria in water from Os and Ve (34.91 and 29.29%, respectively) were higher than those of the other phyla (Figure 9(a)). Bacteroidetes was the most abundant phylum in Ws water, with a relative abundance of 40.76% (Figure 9(a)). All sediment samples in the different groups were mainly occupied by the phylum Proteobacteria, with relative abundance values ranging from 32.66 to 41.71% (Figure 9(b)). The second dominant phylum (relative abundance values >5%, the same as below) in the water of the C was Proteobacteria, followed by Bacteroidetes and Actinobacteria (25.47, 21.98 and 6.08%, respectively, Figure 9(a)). Group C's sediment was Chloroflexi, followed by Firmicutes, Acidobacteria and Bacteroidetes (14.31, 13.02, 12.45 and 5.13%, respectively, Figure 9(b)). The second dominant phylum in the water of Os (25.52, 13.60 and 6.46%, respectively) and Ve (28.03, 15.22 and 6.20%, respectively) was Bacteroidetes, followed by Actinobacteria and Firmicutes (Figure 9(a)). The second dominant phylum in the sediment of Os was Firmicutes, followed by Chloroflexi and Acidobacteria (23.05, 14.05 and 9.41%, respectively, Figure 9(b)). The second dominant phylum in the sediment of Ve was Chloroflexi, followed by Acidobacteria, Firmicutes and Bacteroidetes (21.67, 16.47, 6.71 and 5.56%, respectively, Figure 9(b)). The second dominant phylum in Ws water was Proteobacteria, followed by Firmicutes and Actinobacteria (24.67, 14.25 and 8.98%, respectively, Figure 9(a)). The second dominant phylum in the sediment of Ws was Chloroflexi, followed by Acidobacteria, Bacteroidetes and Firmicutes (23.51, 13.93, 6.16 and 5.96%, respectively, Figure 9(b)). Although Proteobacteria was the most abundant phylum in both water and sediment samples, except water samples from the C and Ws, the relative abundances of this phylum in the sediment from all groups were higher than those in the water. Bacteroidetes, Actinobacteria and Firmicutes were the second dominant phyla in most of the water samples, while Chloroflexi, Acidobacteria, Firmicutes and Bacteroidetes were the second dominant phyla in the sediment samples. Firmicutes, Proteobacteria, Bacteroidetes, Chloroflexi and Actinobacteria can degrade a wide range of substances via anaerobic digestion, which favors the hydrolysis of sewage sludge, and are dominant hydrolytic microbes involved in hydrolysis and anaerobic digestion of sewage sludge (Chouari *et al.* 2005; Chen *et al.* 2017; Peces *et al.* 2018; Ma *et al.* 2019). Consequently, they are always identified in the different types of anaerobic digestion reactors (Ma *et al.* 2019; Shin *et al.* 2019). Previous research revealed that the bacteria in tilapia ponds were dominated by Proteobacteria and Bacteroidetes (Grande Burgos *et al.* 2018; Zhou *et al.* 2018; Liu *et al.* 2020). Increasing the intensity of sewage sludge addition was found to increase the relative abundance of Chloroflexi, Firmicutes and Actinobacteria (Ma *et al.* 2019). The immobilized probiotics changed the structure and diversity of the bacterial community in water and sediment in our study. An enhanced abundance of Firmicutes contributed to better water parameters in group C than other groups. And consistent with lower relative concentrations of NH_4–N, NO_3–N, NO_2–N, TN and COD_{Cr} of immobilized probiotics groups than group C in the sediment, the relative abundance of Firmicutes and Chloroflexi in treated groups were increased. The remediation of the water and sediment might be attributed to the potential ability of these microbes to transfer electrons during the digestion of substances, such as N and organic matter in the water and sediment (Yang *et al.* 2019).

The relative abundances of genus detected with bacterial 16S rRNA sequences in the water and sediment samples in the different groups are shown in Figure 10. The immobilized probiotics changed the relative abundances of the genus in water. *Lactobacillus* was the most abundant genus in the water of the C (Figure 10(a)). The relative abundances of *Flavobacterium* in water from Os, Ve and Ws were higher than the other genus, and *hgcI clade* was the secondary dominant genus in the water of Os and Ve (Figure 10(a)). The relative abundances of *Lactobacillus*, *Escherichia–Shigella* and *Clostridium sensu stricto 1* in water from group C were increased. *Lactobacillus* showed positive impacts on both immunostimulatory and environmental remediation (Talpur *et al.* 2013; Son *et al.* 2018; Flores-Valenzuela *et al.* 2021). In our study, the better water parameters of group C in water than other groups were attributed to the enhanced abundance of *Lactobacillus*. In sediment, *Clostridium sensu stricto 1* and *Escherichia–Shigella* were the dominant genus in the C, Ve and Ws. The application of probiotics immobilized by Os, Ve and Ws changed the microbial community of sediment at the genus level as well. *Lactobacillus*, *Clostridium sensu stricto 1* and *Escherichia–Shigella* were dominant genus in the sediment of the Os (Figure 10(b)). *Escherichia–Shigella*, *Clostridium sensu stricto 1*, *Lactobacillus* and *Haemophilus* of group Os were increased compared with group C in sediment. And remediation effect of group Os on sediment was observed correspondingly. Although the relative abundances of *Escherichia–Shigella*, *Clostridium sensu stricto 1* and *Lactobacillus* of group Ve and Ws were decreased, *Flavobacterium* of group Ve and Ws were increased compared with group C in sediment (Figure 10(b)). Li *et al.* (2012) reported that *Flavobacterium* sp. showed perfect remediation ability of polluted surface water. Therefore, better water parameters of groups Ve and Ws in sediment than group C were attributed to the enhanced abundance of *Flavobacterium*.

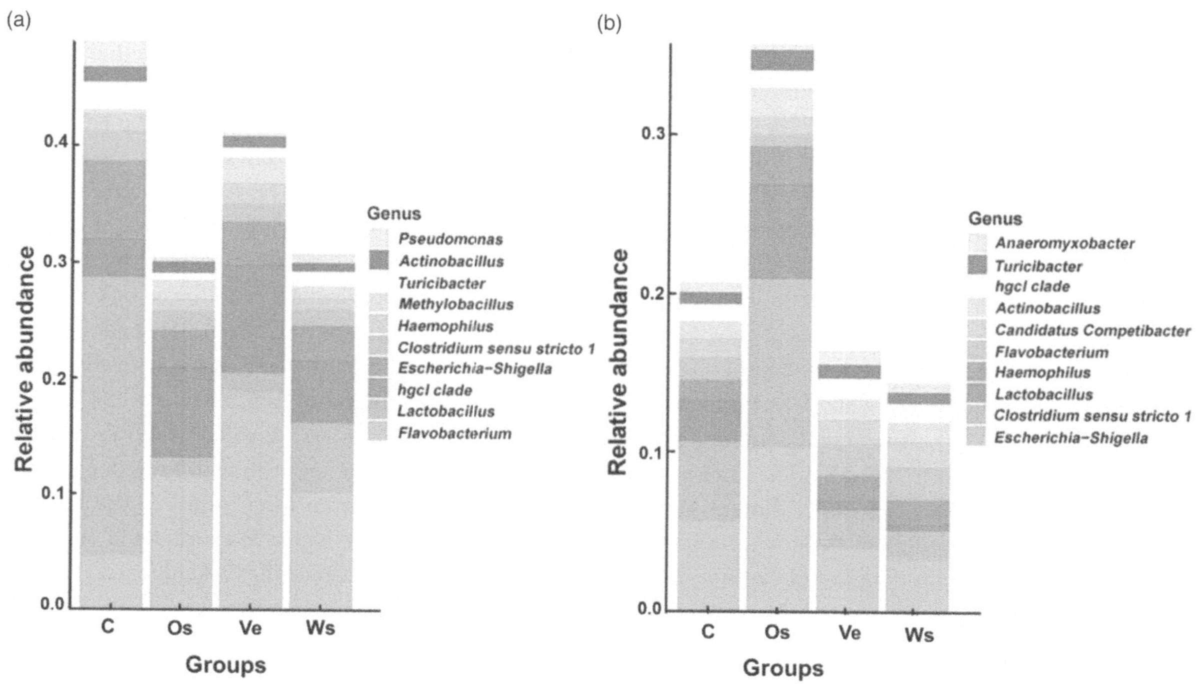

Figure 10 | The relative abundances of microbial community members (genus level) of each group in the water (a) and sediment (b).

CONCLUSION

In the present study, we determined that probiotics immobilized by oyster shells, vesuvianite and walnut shells could remediate aquaculture water and sediment. Oyster shells and walnut shells were effective materials for probiotics immobilization. In addition, the utilization of oyster shells and walnut shells as carriers of immobilized probiotics is a good approach for adding value to these aquaculture and agricultural wastes and increasing the aquaculture water and sediment remediation efficiency of commercial probiotics. It was also found that compared with the effects of immobilized probiotics on water quality parameters, they had better effects on sediment parameters attributed to the changed structure and diversity of the bacterial community. The current research applied immobilized probiotics in an aquaculture pond environment, providing a novel strategy for aquaculture environmental remediation.

ACKNOWLEDGEMENTS

This work was supported by the National Natural Science Foundation of China (No. 31902428), China Agriculture Research System of MOF and MARA (No. CARS-46), Science and Technology Program of Guangzhou, China (No. 201904010304), Key Laboratory of Environment Controlled Aquaculture, Ministry of Education (No. 2021-MOEKLECA-KF-10), Guangdong Basic and Applied Basic Research Foundation (No. 2019A1515111046, 2021A1515010852) and Special Fund for Promoting Economic Development of Guangdong Province (for Modern Fishery Development) (No. YUENONG2019B3, YUENONG2019B8). Mengmeng Yi, Chun Wang and He Wang contributed equally to this work.

DATA AVAILABILITY STATEMENT

All relevant data are included in the paper or its Supplementary Information.

REFERENCES

Asaoka, S., Yamamoto, T., Kondo, S. & Hayakawa, S. 2009 Removal of hydrogen sulfide using crushed oyster shell from pore water to remediate organically enriched coastal marine sediments. *Bioresource Technology* **100** (18), 4127–4132. https://doi.org/10.1016/j.biortech.2009.03.075.

Banerjee, M., Basu, R. K. & Das, S. K. 2018 Cr(VI) adsorption by a green adsorbent walnut shell: adsorption studies, regeneration studies, scale-up design and economic feasibility. *Process Safety and Environmental Protection* **116**, 693–702. https://doi.org/10.1016/j.psep.2018.03.037.

Chang, B. V., Chang, Y. T., Chao, W. L., Yeh, S. L., Kuo, D. L. & Yang, C. W. 2019 Effects of sulfamethoxazole and sulfamethoxazole-degrading bacteria on water quality and microbial communities in milkfish ponds. *Environmental Pollution* **252**, 305–316. https://doi.org/10.1016/j.envpol.2019.05.136.

Chen, Y., Jiang, X., Xiao, K., Shen, N., Zeng, R. J. & Zhou, Y. 2017 Enhanced volatile fatty acids (VFAs) production in a thermophilic fermenter with stepwise pH increase – investigation on dissolved organic matter transformation and microbial community shift. *Water Research* **112**, 261–268. https://doi.org/10.1016/j.watres.2017.01.067.

Chouari, R., Paslier, D. L., Daegelen, P., Ginestet, P., Weissenbach, J. & Sghir, A. 2005 Novel predominant archaeal and bacterial groups revealed by molecular analysis of an anaerobic sludge digester. *Environmental Microbiology* **7** (8), 1104–1115. https://doi.org/10.1111/j.1462-2920.2005.00795.x.

Christianson, L. E., Lepine, C., Sharrer, K. L. & Summerfelt, S. T. 2016 Denitrifying bioreactor clogging potential during wastewater treatment. *Water Research* **105**, 147–156. https://doi.org/10.1016/j.watres.2016.08.067.

Covarrubias, S. A., de-Bashan, L. E., Moreno, M. & Bashan, Y. 2012 Alginate beads provide a beneficial physical barrier against native microorganisms in wastewater treated with immobilized bacteria and microalgae. *Applied Microbiology and Biotechnology* **93**, 2669–2680. https://doi.org/10.1007/s00253-011-3585-8.

Dash, P., Tandel, R. S., Bhat, R. A. H., Mallik, S., Pandey, N. N., Singh, A. K. & Sarma, D. 2018 The addition of probiotic bacteria to microbial floc: water quality, growth, non-specific immune response and disease resistance of *Cyprinus carpio* in mid-Himalayan altitude. *Aquaculture* **495**, 961–969. https://doi.org/10.1016/j.aquaculture.2018.06.056.

FAO 2012 *The State of World Fisheries and Aquaculture 2012 – Meeting the Sustainable Development Goals*. Rome.

FAO 2018 *The State of World Fisheries and Aquaculture 2018 – Meeting the Sustainable Development Goals*. Rome.

Fareed, A., Riaz, S., Nawaz, I., Iqbal, M., Ahmed, R., Hussain, J., Hussain, A., Rashid, A. & Naqvi, T. A. 2019 Immobilized cells of a novel bacterium increased the degradation of *N*-methylated carbamates under low temperature conditions. *Heliyon* **5** (11), e02740. https://doi.org/10.1016/j.heliyon.2019.e02740.

Flores-Valenzuela, E., Miranda-Baeza, A., Rivas-Vega, M. E., Miranda-Arizmendi, V., Beltrán-Ramírez, O. & Emerenciano, M. G. C. 2021 Water quality and productive response of *Litopenaeus vannamei* reared in biofloc with addition of commercial strains of nitrifying bacteria and *Lactobacillus rhamnosus*. *Aquaculture* **542**, 736869. https://doi.org/10.1016/j.aquaculture.2021.736869.

Fu, D., Yan, Y., Yang, X., Rene, E. R. & Singh, R. P. 2020 Bioremediation of contaminated river sediment and overlying water using biologically activated beads: a case study from Shedu river, China. *Biocatalysis and Agricultural Biotechnology* **23**, 101492. https://doi.org/10.1016/j.bcab.2019.101492.

Grande Burgos, M. J., Romero, J. L., Pérez Pulido, R., Cobo Molinos, A., Gálvez, A. & Lucas, R. 2018 Analysis of potential risks from the bacterial communities associated with air-contact surfaces from tilapia (*Oreochromis niloticus*) fish farming. *Environmental Research* **160**, 385–390. https://doi.org/10.1016/j.envres.2017.10.021.

Gregory Caporaso, J., Kuczynski, J., Stombaugh, J., Bittinger, K., Bushman, F. D., Costello, E. K., Fierer, N., Peña, A. G., Goodrich, J. K., Gordon, J. I., Huttley, G. A., Kelley, S. T., Knights, D., Koenig, J. E., Ley, R. E., Lozupone, C. A., McDonald, D., Muegge, B. D., Pirrung, M., Reeder, J., Sevinsky, J. R., Turnbaugh, P. J., Walters, W. A., Widmann, J., Yatsunenko, T., Zaneveld, J. & Knight, R. 2010 QIIME allows analysis of high-throughput community sequencing data. *Nature Methods* **7**, 335–336. https://doi.org/10.1038/nmeth.f.303.

Hasanzadeh, R., Souraki, B. A., Pendashteh, A., Khayati, G. & Ahmadun, F. R. 2020 Application of isolated halophilic microorganisms suspended and immobilized on walnut shell as biocarrier for treatment of oilfield produced water. *Journal of Hazardous Materials* **400**, 123197. https://doi.org/10.1016/j.jhazmat.2020.123197.

Hrenovic, J., Ivankovic, T. & Tibljas, D. 2009 The effect of mineral carrier composition on phosphate-accumulating bacteria immobilization. *Journal of Hazardous Materials* **166** (2–3), 1377–1382. https://doi.org/10.1016/j.jhazmat.2008.12.064.

Jiang, J., Wang, Y., Liu, J., Yang, X., Ren, Y., Miao, H., Pan, Y., Lv, J., Yan, G., Ding, L. & Li, Y. 2019 Exploring the mechanisms of organic matter degradation and methane emission during sewage sludge composting with added vesuvianite: insights into the prediction of microbial metabolic function and enzymatic activity. *Bioresource Technology* **286**, 121397. https://doi.org/10.1016/j.biortech.2019.121397.

Jung, S., Heo, N. S., Kim, E. J., Oh, S. Y., Lee, H. U., Kim, I. T., Hur, J., Lee, G. W., Lee, Y. C. & Huh, Y. S. 2016 Feasibility test of waste oyster shell powder for water treatment. *Process Safety and Environmental Protection* **102**, 129–139. https://doi.org/10.1016/j.psep.2016.03.004.

Kang, A., Mao, H., Li, B., Kou, C., Xu, X. & Jahangiri, B. 2019 Investigation of selective filtration characteristics of filter media for pavement runoff treatment. *Journal of Cleaner Production* **235**, 590–602. https://doi.org/10.1016/j.jclepro.2019.06.337.

Kewcharoen, W. & Srisapoome, P. 2019 Probiotic effects of *Bacillus* spp. from Pacific white shrimp (*Litopenaeus vannamei*) on water quality and shrimp growth, immune responses, and resistance to *Vibrio parahaemolyticus* (AHPND strains). *Fish & Shellfish Immunology* **94**, 175–189. https://doi.org/10.1016/j.fsi.2019.09.013.

Khan, M. D., Ahn, J. W. & Nam, G. 2018 Environmental benign synthesis, characterization and mechanism studies of green calcium hydroxide nano-plates derived from waste oyster shells. *Journal of Environmental Management* **223**, 947–951. https://doi.org/10.1016/j.jenvman.2018.07.011.

Li, H., Ru, J., Yin, W., Liu, X., Wang, J. & Zhang, W. 2009 Removal of phosphate from polluted water by lanthanum doped vesuvianite. *Journal of Hazardous Materials* **168** (1), 326–330. https://doi.org/10.1016/j.jhazmat.2009.02.025.

Li, H. B., Li, Y. H., Wang, H., Wang, X., Zou, T. & Sun, T. H. 2012 Treatment efficiency of polyvinyl alcohol and sodium alginate immobilized *Flavobacterium* sp. *Journal of Shenyang University* **24** (2), 1–5. https://doi.org/10.3969/j.issn.1008-9225.2012.02.002.

Liu, Y. X., Yang, T. O., Yuan, D. X. & Wu, X. Y. 2010 Study of municipal wastewater treatment with oyster shell as biological aerated filter medium. *Desalination* **254** (1–3), 149–153. https://doi.org/10.1016/j.desal.2009.12.003.

Liu, Z., Iqbal, M., Zeng, Z., Lian, Y., Zheng, A., Zhao, M., Li, Z., Wang, G., Li, Z. & Xie, J. 2020 Comparative analysis of microbial community structure in the ponds with different aquaculture model and fish by high-throughput sequencing. *Microbial Pathogenesis* **142**, 104101. https://doi.org/10.1016/j.micpath.2020.104101.

Ma, S. J., Ma, H. J., Hu, H. D. & Ren, H. Q. 2019 Effect of mixing intensity on hydrolysis and acidification of sewage sludge in two-stage anaerobic digestion: characteristics of dissolved organic matter and the key microorganisms. *Water Research* **148**, 359–367. https://doi.org/10.1016/j.watres.2018.10.058.

Ministry of Agriculture and Rural Affairs of the People's Republic of China 2019 *China Fisheries Statistics Yearbook 2019*. China Agricultural Press, Beijing.

Nicholaus, R., Lukwambe, B., Zhao, L., Yang, W., Zhu, J. Y. & Zheng, Z. M. 2019 Bioturbation of blood clam *Tegillarca granosa* on benthic nutrient fluxes and microbial community in an aquaculture wastewater treatment system. *International Biodeterioration & Biodegradation* **142**, 73–82. https://doi.org/10.1016/j.ibiod.2019.05.004.

Nimrat, S., Suksawat, S., Boonthai, T. & Vuthiphandchai, V. 2012 Potential *Bacillus* probiotics enhance bacterial numbers, water quality and growth during early development of white shrimp (*Litopenaeus vannamei*). *Veterinary Microbiology* **159** (3–4), 443–450. https://doi.org/10.1016/j.vetmic.2012.04.029.

Oladoja, N. A., Adelagun, R., Ahmad, A. L. & Ololade, I. A. 2015 Phosphorus recovery from aquaculture wastewater using thermally treated gastropod shell. *Process Safety & Environmental Protection* **98**, 296–308. https://doi.org/10.1016/j.psep.2015.09.006.

Peces, M., Astals, S., Jensen, P. D. & Clarke, W. P. 2018 Deterministic mechanisms define the long-term anaerobic digestion microbiome and its functionality regardless of the initial microbial community. *Water Research* **141**, 366–376. https://doi.org/10.1016/j.watres.2018.05.028.

Pirayesh, H., Khazaeian, A. & Tabarsa, T. 2012 The potential for using walnut (*Juglans regia* L.) shell as a raw material for wood-based particleboard manufacturing. *Composites Part B: Engineering* **43** (8), 3276–3280. https://doi.org/10.1016/j.compositesb.2012.02.016.

Qiu, Z., Chen, J. H., Tang, J. W. & Zhang, Q. Z. 2018 A study of cadmium remediation and mechanisms: improvements in the stability of walnut shell-derived biochar. *Science of the Total Environment* **636**, 80–84. https://doi.org/10.1016/j.scitotenv.2018.04.215.

Schneider, O., Sereti, V., Eding, E. H. & Verreth, J. A. J. 2005 Analysis of nutrient flows in integrated intensive aquaculture systems. *Aquacultural Engineering* **32** (3–4), 379–401. https://doi.org/10.1016/j.aquaeng.2004.09.001.

Schroeder, G. L. 1987 Carbon and nitrogen budgets in manured fish ponds on Israel's coastal plain. *Aquaculture* **62** (3–4), 259–279. https://doi.org/10.1016/0044-8486(87)90169-4.

Segovia-Sandoval, S. J., Ocampo-Pérez, R., Berber-Mendoza, M. S., Leyva-Ramos, R., Jacobo-Azuara, A. & Medellín-Castillo, N. A. 2018 Walnut shell treated with citric acid and its application as biosorbent in the removal of Zn(II). *Journal of Water Process Engineering* **25**, 45–53. https://doi.org/10.1016/j.jwpe.2018.06.007.

Shih, P. K. & Chang, W. L. 2015 The effect of water purification by oyster shell contact bed. *Ecological Engineering* **77**, 382–390. https://doi.org/10.1016/j.ecoleng.2015.01.014.

Shin, J., Cho, S. K., Lee, J., Hwang, K., Chung, J. W., Jang, H. N. & Shin, S. G. 2019 Performance and microbial community dynamics in anaerobic digestion of waste activated sludge: impact of immigration. *Energies* **12**, 573. https://doi.org/10.3390/en12030573.

Son, S. H., Yang, S. J., Jeon, H. L., Yu, H. S., Lee, N. K., Park, Y. S. & Paik, H. D. 2018 Antioxidant and immunostimulatory effect of potential probiotic *Lactobacillus paraplantarum* SC61 isolated from Korean traditional fermented food, jangajji. *Microbial Pathogenesis* **125**, 486–492. https://doi.org/10.1016/j.micpath.2018.10.018.

Song, X., Li, K., Wang, C., Sun, X., Ning, P. & Tang, L. 2017 Regeneration performance and mechanism of modified walnut shell biochar catalyst for low temperature catalytic hydrolysis of organic sulfur. *Chemical Engineering Journal* **330**, 727–735. https://doi.org/10.1016/j.cej.2017.08.016.

Talpur, A. D., Ikhwanuddin, M., Abdullah, M. D. D. & Bolong, A.-M. A. 2013 Indigenous *Lactobacillus plantarum* as probiotic for larviculture of blue swimming crab, *Portunus pelagicus* (Linnaeus, 1758): effects on survival, digestive enzyme activities and water quality. *Aquaculture* **416–417**, 173–178. https://doi.org/10.1016/j.aquaculture.2013.09.018.

Wang, C., Gao, J. & Gu, C. 2017 Rapid destruction of tetrabromobisphenol A by iron(III)-tetraamidomacrocyclic ligand/layered double hydroxide composite/H_2O_2 system. *Environmental Science & Technology* **51** (1), 488–496. https://doi.org/10.1021/acs.est.6b04294.

Wójcik, M. 2020 Investigation of filtration properties and microbiological characteristics of sewage sludge after physical conditioning with the use of ground walnut shells. *Powder Technology* **361**, 491–498. https://doi.org/10.1016/j.powtec.2019.11.100.

Wu, Q., Chen, J., Clark, M. & Yu, Y. 2014 Adsorption of copper to different biogenic oyster shell structures. *Applied Surface Science* **311**, 264–272. http://dx.doi.org/10.1016/j.apsusc.2014.05.054.

Xia, L., Li, X., Fan, W. & Wang, J. 2020 Heterotrophic nitrification and aerobic denitrification by a novel *Acinetobacter* sp. ND7 isolated from municipal activated sludge. *Bioresource Technology* **301**, 122749. https://doi.org/10.1016/j.biortech.2020.122749.

Xue, J. L., Wu, Y. N., Liu, Z. X., Li, M. L., Sun, X. Y., Wang, H. J. & Liu, B. 2017 Characteristic assessment of diesel-degrading bacteria immobilized on natural organic carriers in marine environment: the degradation activity and nutrient. *Scientific Reports* **7** (1), 8635. https://doi.org/10.1038/s41598-017-08832-y.

Yang, G., Wang, J., Zhang, H., Jia, H., Zhang, Y. & Gao, F. 2019 Applying bio-electric field of microbial fuel cell-upflow anaerobic sludge blanket reactor catalyzed blast furnace dusting ash for promoting anaerobic digestion. *Water Research* **149**, 215–224. https://doi.org/10.1016/j.watres.2018.10.091.

Yen, H. Y. & Chou, J. H. 2016 Water purification by oyster shell bio-medium in a recirculating aquaponic system. *Ecological Engineering* **95**, 229–236. https://doi.org/10.1016/j.ecoleng.2016.06.065.

Zare, H., Najafpour, G., Rahimnejad, M., Tardast, A. & Gilani, S. 2012 Biofiltration of ethyl acetate by *Pseudomonas putida* immobilized on walnut shell. *Bioresource Technology* **123**, 419–423. https://doi.org/10.1016/j.biortech.2012.07.036.

Zhou, M., Wan, Q. Y., Sarath Babu, V., Qiu, Q., Kou, H. Y., Lin, C., Zhao, L. J., Yang, L., Li, J., Huang, Y. M. & Lin, L. 2018 Bacterial features in tilapia (*Oreochromis niloticus*) and environments in a goose-tilapia polyculture model. *Aquaculture* **497**, 313–319. https://doi.org/10.1016/j.aquaculture.2018.07.063.

Zhu, G. L., Hu, Y. Y. & Wang, Q. R. 2009 Nitrogen removal performance of anaerobic ammonia oxidation co-culture immobilized in different gel carriers. *Water Science & Technology* **59** (12), 2379–2386. https://doi.org/10.2166/wst.2009.293.

Zhu, B., Chen, S., Zhao, C., Zhong, W., Zeng, R. & Yang, S. 2019 Effects of *Marichromatium gracile* YL28 on the nitrogen management in the aquaculture pond water. *Bioresource Technology* **292**, 121917. https://doi.org/10.1016/j.biortech.2019.121917.

Zokaeifar, H., Babaei, N., Che, R. S., Kamarudin, M. S., Sijam, K. & Balcazar, J. L. 2014 Administration of *Bacillus subtilis* strains in the rearing water enhances the water quality, growth performance, immune response, and resistance against *Vibrio harveyi* infection in juvenile white shrimp, *Litopenaeus vannamei*. *Fish & Shellfish Immunology* **36** (1), 68–74. https://doi.org/10.1016/j.fsi.2013.10.007.

First received 10 June 2021; accepted in revised form 13 September 2021. Available online 28 September 2021

© 2022 The Authors

Water Recycling and Low-Carbon Development doi: 10.2166/wrd.2021.063

Effect of powder-activated carbon pre-coating membrane on the performance of the UF system for wastewater reclamation: a pilot-scale study

Tong Yu, Haoshuai Yin, Lihua Cheng and Xuejun Bi*

School of Environmental and Municipal Engineering, Qingdao University of Technology, Qingdao 266000, China
*Corresponding author. E-mail: xuejunb@126.com

LC, 0000-0002-4806-5239

ABSTRACT

Pretreatment is an indispensable means to alleviate membrane fouling and improve ultrafiltration (UF) performance. In this study, we designed and established a powder-activated carbon pre-coating membrane (PACPM) unit as a pretreatment for the UF system. The effective filtration area of PACPM was 0.5 m^2. 300 g/m^2 PAC was selected as the optimal dosage in the pilot-scale apparatus according to the laboratory-scale trial. The pre-coating membrane could be formed within 30 min. PACPM could remove different kinds of organic compounds with different molecular weights and hydrophobicity during a certain period of time. During 10 operation cycles, the specific flux decrease rate of PACPM-UF was only 2.8%, which was much lower than that of direct UF (13.4%). PACPM could improve the performance of the UF system, not only for the increase of the initial SF value but also for the increase of the flux recovery thorough backwash. Nevertheless, a regular replace-regeneration process is necessary to maintain PACPM performance.

Key words: membrane fouling, pilot-scale system, powder-activated carbon pre-coating membrane (PACPM), UF membrane

HIGHLIGHTS

- The pilot-scale powder-activated carbon pre-coating membrane (PACPM) unit was designed and established.
- The pre-coating membrane could be formed in a short time.
- PACPM-UF had a better efficiency of contaminant removal than direct UF.
- PACPM could improve the performance of the UF system treating the municipal wastewater treatment plant effluent.

GRAPHICAL ABSTRACT

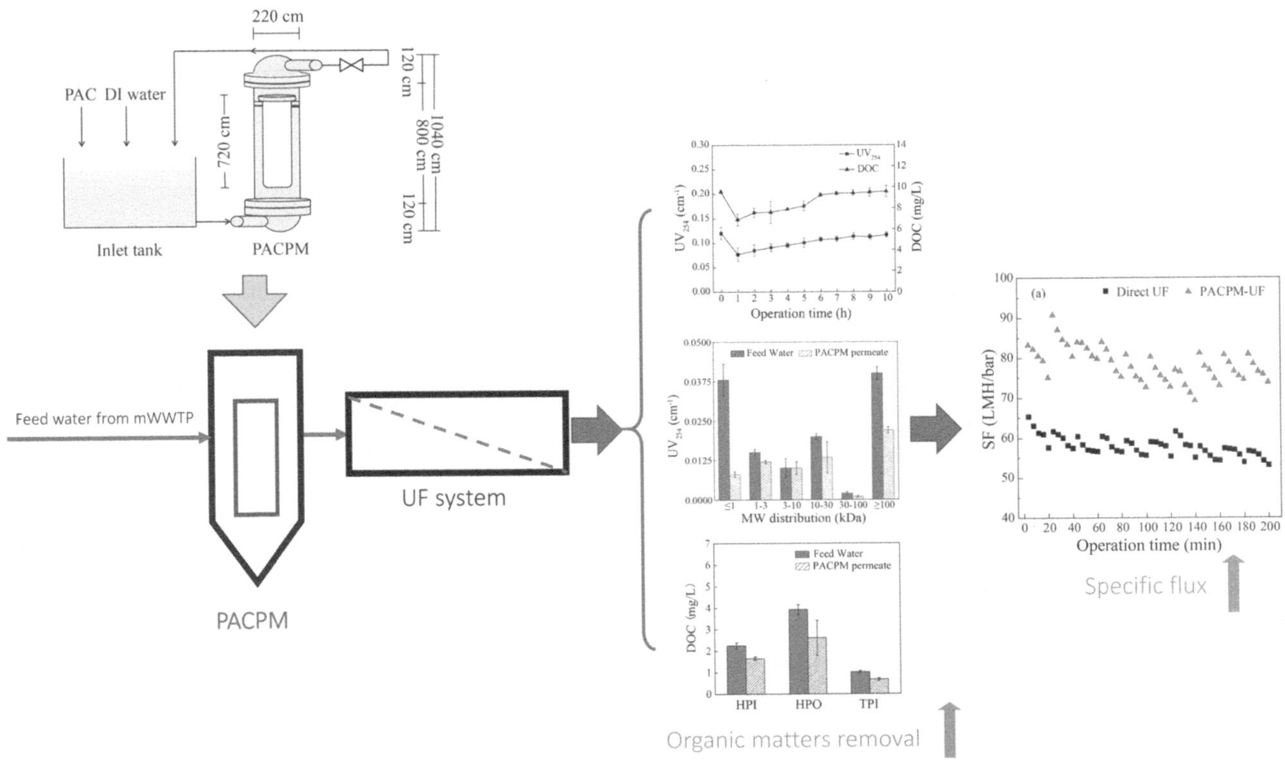

INTRODUCTION

Wastewater reclamation is an attractive approach for overcoming the global water crisis (Nahrstedt *et al.* 2020). Ultrafiltration (UF) technology has become increasingly applied for wastewater reclamation and leachate treatment due to the stable permeate quality and automatic operation (Snyder *et al.* 2007; Qu *et al.* 2015; Qu *et al.* 2019; Abuabdou *et al.* 2021; Yu *et al.* 2021). However, organic fouling of UF membranes is still a major impediment to the effective operation of UF systems. Organic fouling causes multiple adverse effects on the performance of UF systems (e.g., decreased production, water quality deterioration and decreased module lifetime) (Zhao *et al.* 2018; Ahmad *et al.* 2020).

Pretreatment can increase the permeate quality and reduce UF membrane fouling (Huang *et al.* 2009; Wang *et al.* 2020). Powder-activated carbon (PAC) absorption is one of the most popular pretreatment techniques because of its contaminant adsorption ability and commercial availability (Gao *et al.* 2011). Although much work has been done to study the effect of PAC on UF membrane fouling, the results are somehow contradictory: some researchers found that PAC could not only remove organic matter but also alleviate UF membrane fouling (Campinas & Rosa 2010; Sun *et al.* 2017; Cheng *et al.* 2021), while other researchers reported that PAC could not alleviate UF membrane fouling (Lee *et al.* 2000; Zhang *et al.* 2003; Shao *et al.* 2016; Zhang *et al.* 2018).

Pre-coating membrane is a filter cake layer formed by pre-coated agent on the surface of base membrane, which could improve the filtration accuracy and reduce the fouling of the base membrane (Ma *et al.* 2013; Yang *et al.* 2015). However, it was reported that PAC directly coating on the UF membrane was liable to cause UF membrane damage and increase backwash difficulty (Malczewska *et al.* 2015). Compared with the direct coating on the UF membrane, PAC pre-coating on the base membrane could avoid the damage to the UF membrane. The PAC cake layer and the base membrane could not only adsorb organic matter but also remove colloids and particles in the influent, which could protect the UF membrane from damage. However, to date there are few reports on the powder-activated carbon pre-coating membrane (PACPM) as the UF system pretreatment for wastewater reclamation, especially in pilot-scale studies.

In this study, a pretreatment technique using PAC pre-coating on the base membrane was investigated to improve the performance of the UF system. The dosage and organic matter removal efficiency of PAC were determined via laboratory-scale experiments. The pilot-scale PACPM-UF system was set up to investigate the effects of PACPM on contaminant removal and the performance of the UF system treating the effluent of a municipal wastewater treatment plant (mWWTP).

METHODS

Water samples

PACPM feed water, lab-scale and pilot-scale PACPM permeates were collected and stored at 4 °C prior to analyses. The feed water was collected from the effluent of a coastal municipal wastewater treatment plant (cmWWTP), which was located in Qingdao, Shandong Province, China. The cmWWTP adopted anaerobic–anoxic–oxic with moving-bed biofilm reactor as a secondary treatment process for the biological removal of nitrogen and phosphorus, with additional coagulation, filtration and UV disinfection as advanced treatment. Feed water characteristics are listed in Table 1. The conductivity was 1,810–9,790 μS/cm, which was significantly higher than other common mWWTPs (Xu *et al.* 2010; Chon *et al.* 2012; Ayache *et al.* 2013; Wang *et al.* 2019; Tong *et al.* 2020), which could result from the periodical seawater intrusion of this cmWWTP.

Laboratory-scale experimental setup

A vacuum filter holder was applied for the lab-scale experiment to determine the dosage and organic matter removal efficiency (Supplementary Figure S1). The qualitative filter paper was put on the core of the filter as the base membrane. The filtering area was 1.67×10^{-3} m^2. The filtration pressure was maintained at 0.07 MPa via a vacuum pump. PAC with an average particle diameter of 32 μm and an average pore size of 1 nm was used to form the pre-coating dynamic membrane (purchased from Sinopharm Chemical Reagent Co., Ltd). To fabricate lab-scale PACPM, different loads of PAC (0, 150, 300, 600, 900 1,200 and 1,500 g/m^2) were suspended by deionized (DI) water and filtered by vacuum filtration onto a qualitative filter paper. The feed water sample (500 mL) was pumped through PACPM at 0.07 MPa using a vacuum pump. The flux of the feed water sample was calculated by time recording. Then, PACPM permeate was pumped through the 0.2-μm nylon membrane (Whatman, England) at 0.07 MPa. The flux of PACPM permeate was calculated by time recording.

Pilot-scale experimental setup

The pilot-scale PACPM apparatus was designed and fabricated based on the bag filter, schematically shown in Figure 1. The effective filtration area of the base membrane was 0.5 m^2. To fabricate PACPM, PAC particles were suspended with DI water into the PACPM feed tank. Then, the PAC suspension was pumped into the bag filter and circulated in the PACPM apparatus. The PACPM-UF system, as shown in Figure 2, has a capacity of 72 m^3/d. The characteristics of the UF membrane (inge dizzer® 5000 plus 0.9MB50) used in the pilot system are shown in Table 2. The water backwash of the UF membrane was performed with a flux of 230 L/(m^2·h) and a duration of 40 s. The chemical enhanced backwash (CEB) process was performed on a daily basis. The sequence of the CEB process was as follows: alkaline cleaning (NaOH, pH 12, 40 s), alkaline immersion (10–60 min), water backwash (40 s), acid cleaning (H_2SO_4, pH 2, 40 s), acid immersion (40 min), and then water backwash (40 s). The UF permeate was used for the water backwash and CEB process.

Table 1 | Feed water characteristics

Characteristics	Concentration range	Average value
pH	7.05–7.87	7.46
SS (mg/L)	1–9	4
Conductivity (μS/cm)	1,810–9,790	5,500
Turbidity (NTU)	0.53–2.93	1.13
DOC (mg/L)	7.68–11.35	9.62
UV_{254} (cm^{-1})	0.10–0.13	0.12
TN (mg/L)	7.94–14.30	11.12

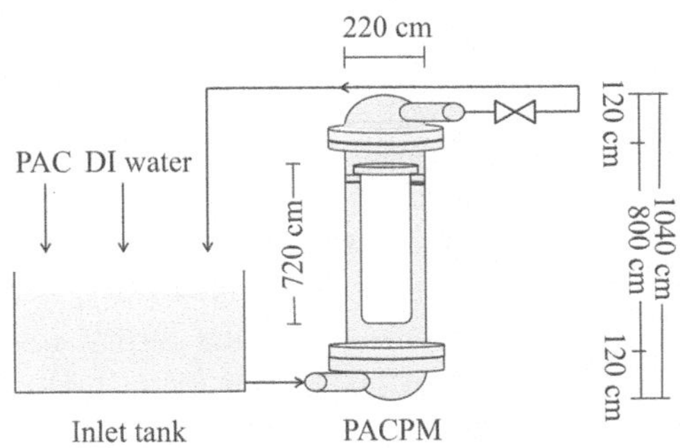

Figure 1 | Schematic diagram of the pilot-scale PACPM apparatus.

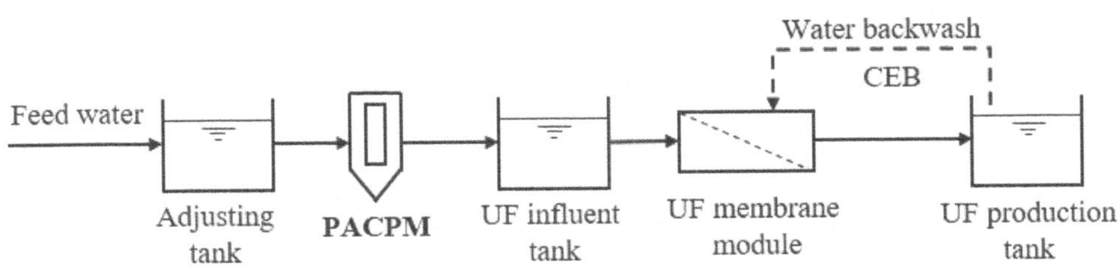

Figure 2 | Schematic diagram of the major process in the pilot PACPM-UF system.

Table 2 | Characteristics of the UF membrane applied in the process

Items	UF membrane
Manufacturer	inge dizzer®
Type	5000 plus 0.9MB50
Membrane material	PESM
Pore size (μm)	0.02
Effective surface area (m^2)	50
Filtration mode	Dead-end
Membrane flux ($L/(m^2{\cdot}h)$)	60
maximum pressure (MPa)	0.5
Recovery rate (%)	90

Analytical methods

Transmembrane pressure (TMP) and membrane flux (J) of the UF membrane were measured by the online monitor of the PACPM-UF system. Specific flux (SF) was defined as membrane flux (J) divided by TMP. pH and conductivity were measured by a multi-parameter controller (LEICI, China). Suspended solids (SS) was measured by an ultraviolet-visible spectrophotometer (HACH, USA). Turbidity was measured by an SGZ-20B portable turbidimeter (Mingbolm, China). Dissolved organic carbon (DOC) and total nitrogen (TN) were measured by a Multi N/C 2100 analyzer (Analytik Jena, Germany). UV absorbance at 254 nm (UV_{254}) was measured by a Uvmini-1240 spectrophotometer (Shimadzu, Japan). The molecular weight (MW) distribution of the organic matter (UV_{254}) in the water sample was measured using cellulose ultrafiltration membranes with MW cut-offs of 1, 3, 10, 30 and 100 kDa (Millipore, USA) as described by Zhao *et al.* (2014). Resin fractionations of the dissolved organic matter (DOM) in the water samples were performed using DAX-8 and XAD-4 ion exchange resins with a method described by Zularisam *et al.* (2007).

RESULTS AND DISCUSSION

Optimal PAC dosage of PACPM

To determine the optimal PAC dosage of PACPM, a lab-scale apparatus was applied using different loads of PAC. With the increase of PAC dosage, the flux of the feed water through PACPM decreased rapidly from 5.4 to 3.1 m/h at the beginning and then decreased steadily (Figure 3(a)). Correspondingly, the flux of PACPM permeate through the 0.2-μm membrane increased from 3.8 to 7.1 m/h (Figure 3(b)). Regarding the removal of the organic matter, with the increase of PAC dosage from 0 to 300 g/m^2, the DOC and UV_{254} removal rates increased to 41.7 and 89.5%, respectively. Subsequently, the DOC and UV_{254} removal rates kept stable with the increase of PAC dosage from 300 to 1,500 g/m^2 (Figure 3(c) and 3(d)). Considering the effects of the PAC dosage on both flux and organics removal, 300 g/m^2 was selected as the PAC dosage in the pilot-scale study.

Effects of PACPM on contaminant removal

As the effective filtration area of the base membrane was 0.5 m^2, 150 g PAC (300 g/m^2) particles were suspended with DI water and pumped from the bottom of the PACPM apparatus. The PAC particles gradually and evenly adhered to the base membrane with the circulation of the PAC suspension in the apparatus (Figure 4). Meanwhile, PACPM permeate became more and more clear (Supplementary Figure S2). After 30 min circulation of the PAC suspension, the SS of the permeate decreased to less than 5 mg/L and remained stable, which indicated that the pre-coating membrane has been successfully developed (Figure 5).

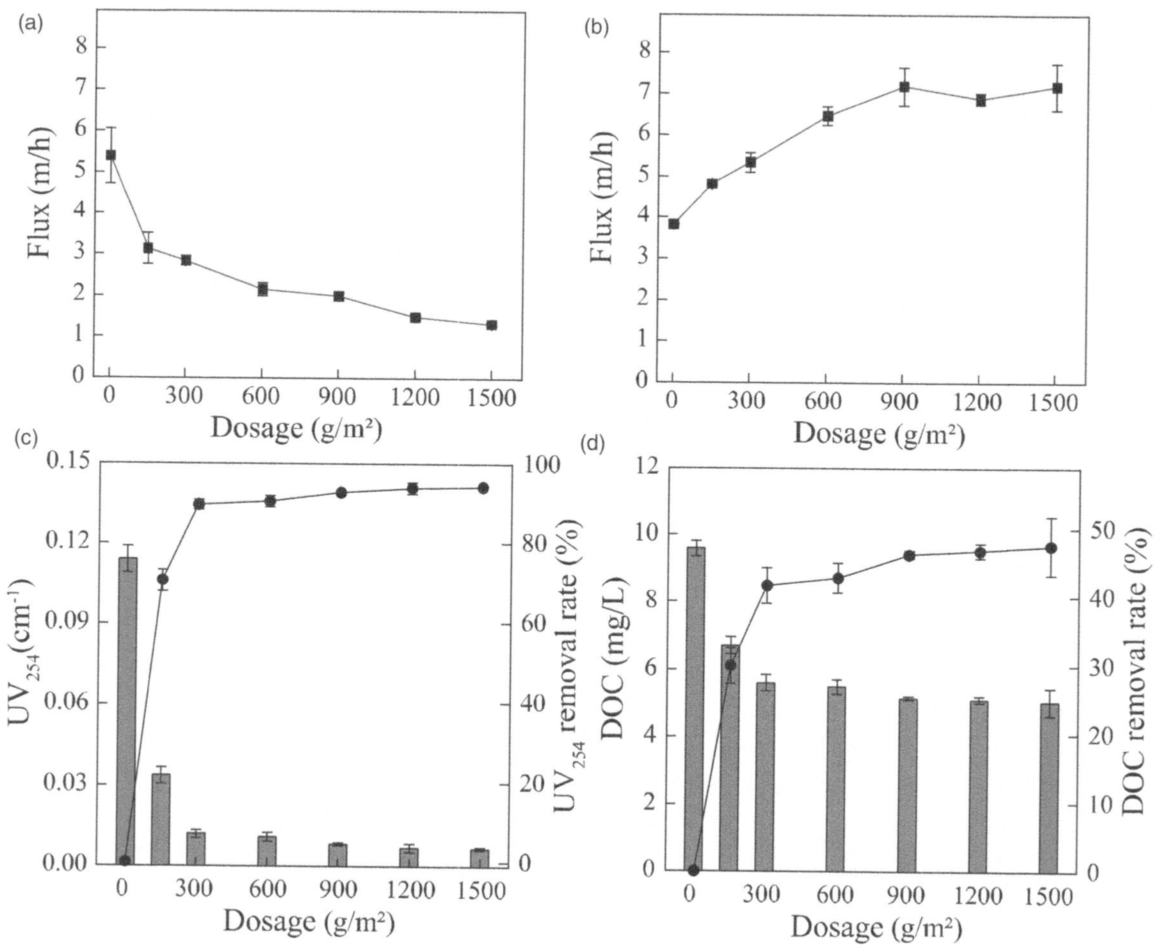

Figure 3 | Effect of PACPM dosages on the flux and organics removal. (a) Flux of the feed water sample through the PACPM. (b) Flux of the PACPM permeate through the 0.2-μm membrane (c) UV_{254} removal and (d) DOC removal.

Figure 4 | Coating process of the pilot-scale PACPM.

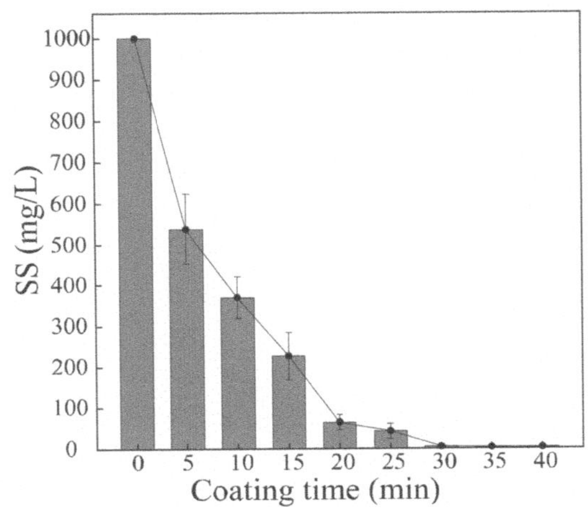

Figure 5 | Changes of SS of PACPM permeate during the coating process.

With the accomplishment of the coating progress, the feed water from the cmWWTP was treated continuously by PACPM at 0.02 MPa (Figure 2). The DOC and UV_{254} of PACPM permeate were measured hourly to evaluate the contaminant removal capacity of PACPM. As shown in Figure 6, the concentrations of the DOC and UV_{254} decreased sharply from 9.55 to 6.91 mg/L and from 0.120 to 0.077 cm^{-1} in the first hour, respectively. The removal rates of DOC and UV_{254} in the first

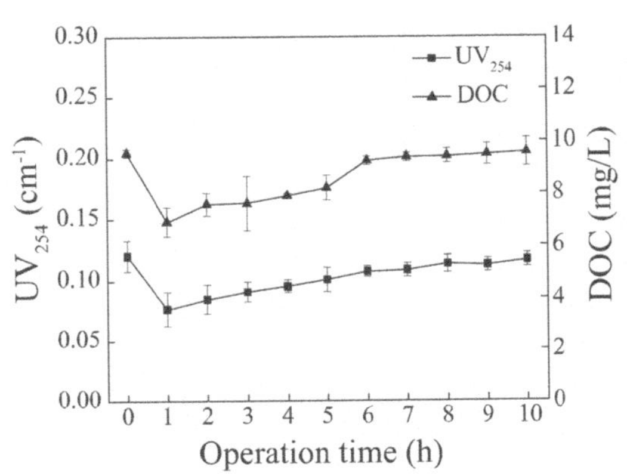

Figure 6 | Removal of DOC and UV_{254} by PACPM in different operation times.

hour were 27.6 and 35.8%. However, with the increase of operation time, the contaminant accumulated in the PAC particles, which led to a gradual decrease in the removal efficiency of organic matter by PAC. After 8 h of operation, the removal rate of UV_{254} by PACPM was reduced to about 5%, while the removal rate of DOC by PACPM was nearly zero. These results indicated that PACPM could effectively remove organic matter within a certain range, but it would gradually lose its removal capacity, which was consistent with the absorption characteristics of PAC.

The MW distributions of the organic matter of feed water and PACPM permeate in the first hour were measured to investigate the capability of PACPM to remove contaminants with different MWs. The result indicated that most of the organics with different MWs could be removed by PACPM. Among them, PACPM showed better removal ability for organic matter less than 1 kDa and more than 100 kDa, which could be explained by a combined effect of adsorption and filtration (Figure 7).

Similarly, the DOM fractions of feed water and PACPM permeate in the first hour were measured to investigate the capability of PACPM to remove different DOM fractional components. The water samples were fractionated into several components which were hydrophilic (HPI), hydrophobic (HPO) and transphilic (TPI) fractions using DAX-8 and XAD-4 ion exchange resins. The concentrations of HPI, HPO and TPI in the feed water were 2.24, 3.93 and 1.01 mg/L, respectively. After PACPM treatment, the concentrations of HPI, HPO and TPI decreased to 1.6, 2.61 and 0.67 mg/L, respectively. The removal rates of HPI, HPO and TPI were 26.34, 33.59 and 33.66%, respectively (Figure 8). The result indicated that both hydrophilic and hydrophobic fractions could be removed by PACPM, and the removal capacity of hydrophobic fraction by PACPM was slightly higher than that of hydrophilic fraction. The result was consistent with another study evaluating the removal of organic fraction by PAC (Wang *et al.* 2016).

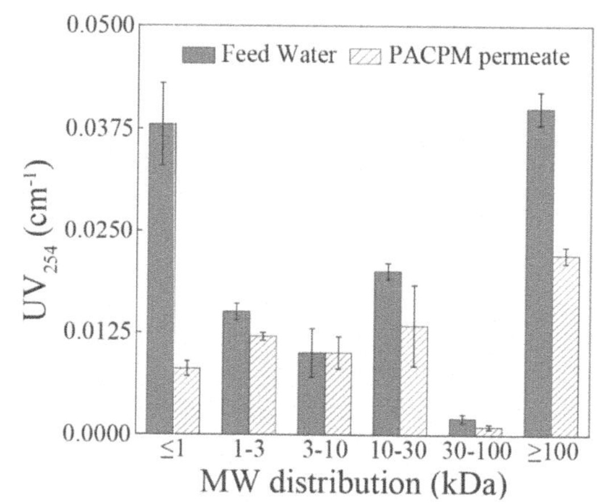

Figure 7 | MW distributions of feed water and PACPM permeate.

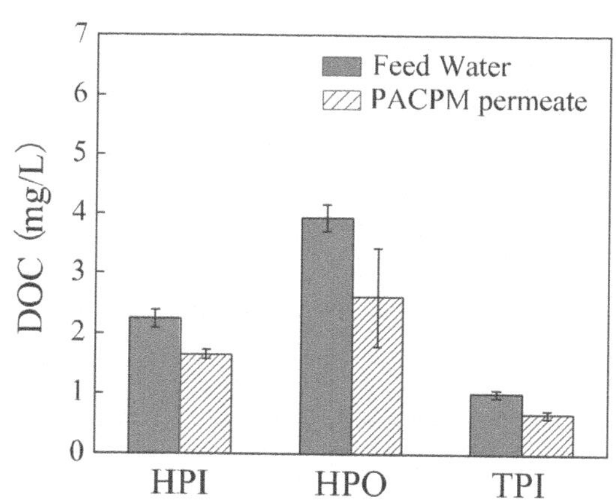

Figure 8 | DOM fractions of feed water and PACPM permeate.

Effects of PACPM on the performance of the UF system

To evaluate the effects of PACPM on the performance of the UF system, the SF variations between direct UF and PACPM-UF were compared (Figure 9). One operation cycle lasted for 20 min. The water backwash of the UF membrane was performed every 20 min operation with a duration of 40 s. The initial SF value of PACPM-UF was 83.26 LMH/bar, which was much higher than that of direct UF (65.37 LMH/bar). As shown in Figure 9(a), during 10 operation cycles, the SF values of PACPM-UF were always significantly higher than those of direct UF. Within each cycle, the decrease of SF for PACPM-UF was higher than direct UF. However, the recovery by water backwash of PACPM-UF was much better than that of direct UF (Figure 9(a)). After 10 operation cycles, the SF value of direct UF decreased from 65.37 to 56.64 LMH/bar, and the decrease rate was 13.4%. However, the SF value of PACPM-UF only decreased from 83.26 to 80.89 LMH/bar, and the decrease rate was only 2.8% (Figure 9(b)). These results indicated that PACPM could improve the performance of the UF system in a certain period of operation time, not only for the increase of the initial SF value but also for the increase of the recovery by water backwash.

The 24-h continuous operation experiments were conducted to assess the durability of PACPM for improving the performance of the UF system. As shown in Figure 10, the SF values of PACPM-UF were significantly higher than those of direct UF

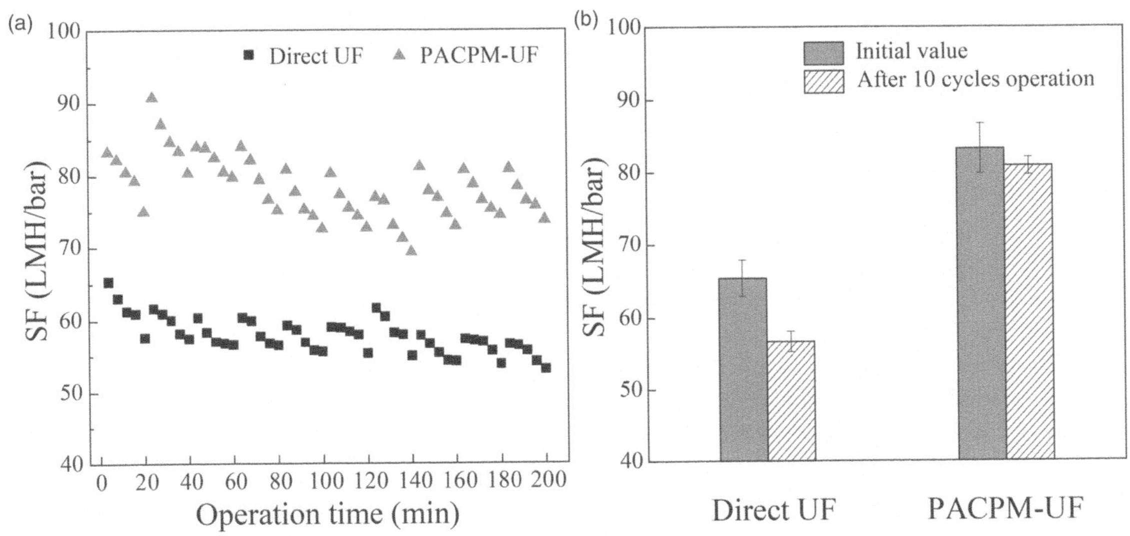

Figure 9 | Comparison of SF variations between direct UF and PACPM-UF: (a) SF variations in the period of 10 operation cycles and (b) SF of initial value and the value after 10 operation cycles.

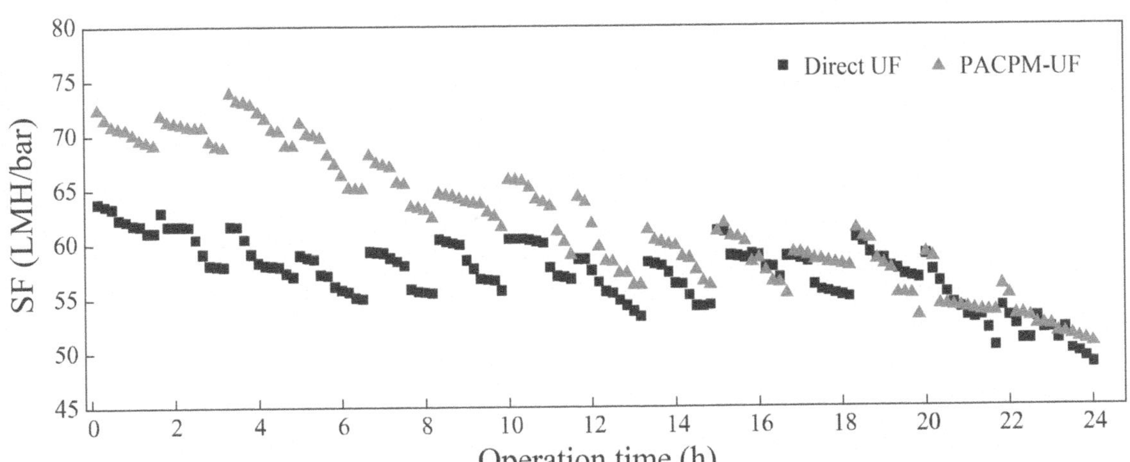

Figure 10 | Comparison of SF variations between direct UF and PACPM-UF in 24 h.

Table 3 | Water qualities of water backwash effluents

	Direct UF	PACPM-UF
UV_{254} (cm^{-1})	0.167 ± 0.015	0.114 ± 0.017
DOC (mg/L)	8.41 ± 0.33	7.31 ± 0.58
Turbidity (NTU)	21.39 ± 10.61	6.16 ± 0.96
SS (mg/L)	19.67 ± 15.80	13.12 ± 5.61

Table 4 | Water qualities of alkaline cleaning effluents of the CEB process

	Direct UF	PACPM-UF
UV_{254} (cm^{-1})	1.57 ± 0.69	0.79 ± 0.11
DOC (mg/L)	76.72 ± 8.82	44.18 ± 3.61
Turbidity (NTU)	7.48 ± 6.45	5.83 ± 3.28
SS (mg/L)	5.83 ± 2.20	6.01 ± 2.83

during the first 12 h. Even though PACPM was almost unable to remove organic matter after 8-h operation (Figure 6), the improvement of UF operation was maintained up to more than 12 h (Figure 10). This may be due to the formed pre-coating membrane which could still remove some particles and colloids. After 12 h, the performance of PACPM deteriorated further and the membrane itself became clogged, resulting in the SF value of PACPM-UF gradually approaching that of direct UF. The result suggested that PACPM could only improve the performance of the UF system for a period of time. Therefore, in order to improve the performance of the UF system steadily, other means may need to be implemented, such as increasing the dosage of PAC, increasing the filtration area, parallel operation of multiple groups of PACPM, regular replacement and regeneration of PAC.

To investigate the effect of PACPM on the UF membrane fouling, the water qualities of effluents for water backwash and the CEB process were analyzed. As shown in Table 3, the organic matter and particles in the backwash water for PACPM-UF were lower than those in the backwash water for direct UF. Similarly, the organic matter in the CEB alkaline cleaning effluent of PACPM-UF was also significantly lower than that of direct UF (Table 4). The water quality of backwash and CEB alkaline cleaning effluents showed that PACPM could effectively reduce the contaminant accumulation on the UF membrane. The recovery by the water backwash indicated that PACPM could remove the contaminant which was more difficult to rinse out from the UF membrane surface so as to improve the effect of water backwash (Figure 9).

CONCLUSIONS

In this work, we designed and established the PACPM with a bag filter as the base membrane. Based on the results of lab-scale experiments, the optimal PAC dosage of PACPM was determined as 300 g/m^2. The results of pilot-scale experiments indicated that PACPM could remove different kinds of organic compounds with different MWs and hydrophobicity during a certain period of time. Furthermore, PACPM could improve the performance of the UF system during a certain period of operation time, not only for the increase of the initial SF value but also for the increase of the recovery by water backwash. After PACPM lost the ability of organic matter removal, it could still maintain the optimization effect on the UF system for a certain period of time. Nevertheless, a regular replace-regeneration process is necessary to maintain PACPM performance.

ACKNOWLEDGEMENTS

This study was supported by the China Postdoctoral Science Foundation (No. 2019M652347), the Major Science and Technology Program for Water Pollution Control and Treatment (No. 2017ZX07101002-06), and College Innovative Research Team of Shandong Province (2020KJD003).

DATA AVAILABILITY STATEMENT

All relevant data are included in the paper or its Supplementary Information.

REFERENCES

Abuabdou, S. M. A., Jaffari, Z. H., Ng, C.-A., Ho, Y.-C. & Bashir, M. J. K. 2021 A new polyvinylidene fluoride membrane synthesized by integrating of powdered activated carbon for treatment of stabilized leachate. *Water* **13** (16), 2282.

Ahmad, M. A., Zainal, B. S., Jamadon, N. H., Yaw, T. & Abdullah, L. C. 2020 Filtration analysis and fouling mechanisms of PVDF membrane for POME treatment. *Water Reuse* **10** (3), 187–199.

Ayache, C., Manes, C., Pidou, M., Croue, J.-P. & Gernjak, W. 2013 Microbial community analysis of fouled reverse osmosis membranes used in water recycling. *Water Research* **47** (10), 3291–3299.

Campinas, M. & Rosa, M. J. 2010 Assessing PAC contribution to the NOM fouling control in PAC/UF systems. *Water Research* **44** (5), 1636–1644.

Cheng, X., Hou, C., Li, P., Luo, C., Zhu, X., Wu, D., Zhang, X. & Liang, H. 2021 The role of PAC adsorption-catalytic oxidation in the ultrafiltration performance for treating natural water: efficiency improvement, fouling mitigation and mechanisms. *Chemosphere* **284**, 131561.

Chon, K., Cho, J., Shon, H. K. & Chon, K. 2012 Advanced characterization of organic foulants of ultrafiltration and reverse osmosis from water reclamation. *Desalination* **301**, 59–66.

Gao, W., Liang, H., Ma, J., Han, M., Chen, Z.-l., Han, Z.-s. & Li, G.-b. 2011 Membrane fouling control in ultrafiltration technology for drinking water production: a review. *Desalination* **272** (1), 1–8.

Huang, H., Schwab, K. & Jacangelo, J. G. 2009 Pretreatment for low pressure membranes in water treatment: a review. *Environmental Science & Technology* **43** (9), 3011–3019.

Lee, S.-J., Choo, K.-H. & Lee, C.-H. 2000 Conjunctive use of ultrafiltration with powdered activated carbon adsorption for removal of synthetic and natural organic matter. *Journal of Industrial and Engineering Chemistry* **6** (6), 357–364.

Ma, J., Wang, Z., Xu, Y., Wang, Q. & Grasmick, A. 2013 Organic matter recovery from municipal wastewater by using dynamic membrane separation process. *Chemical Engineering Journal* **219**, 190–199.

Malczewska, B., Liu, J. & Benjamin, M. M. 2015 Virtual elimination of MF and UF fouling by adsorptive pre-coat filtration. *Journal of Membrane Science* **479**, 159–164.

Nahrstedt, A., Gaba, A., Zimmermann, B., Jentzsch, T. & Rohn, A. 2020 Reuse of municipal wastewater for different purposes based on a modular treatment concept. *Water Reuse* **10** (4), 301–316.

Qu, F., Yan, Z., Liu, W., Shao, S., Ren, X., Ren, N., Li, G. & Liang, H. 2015 Effects of manganese dioxides on the ultrafiltration membrane fouling by algal extracellular organic matter. *Separation and Purification Technology* **153**, 29–36.

Qu, F., Wang, H., He, J., Fan, G., Pan, Z., Tian, J., Rong, H., Li, G. & Yu, H. 2019 Tertiary treatment of secondary effluent using ultrafiltration for wastewater reuse: correlating membrane fouling with rejection of effluent organic matter and hydrophobic pharmaceuticals. *Environmental Science: Water Research Technology* **5** (4), 672–683.

Shao, S., Liang, H., Qu, F., Li, K., Chang, H., Yu, H. & Li, G. 2016 Combined influence by humic acid (HA) and powdered activated carbon (PAC) particles on ultrafiltration membrane fouling. *Journal of Membrane Science* **500**, 99–105.

Snyder, S. A., Adham, S., Redding, A. M., Cannon, F. S., Decarolis, J., Oppenheimer, J., Wert, E. C. & Yoon, Y. 2007 Role of membranes and activated carbon in the removal of endocrine disruptors and pharmaceuticals. *Desalination* **202** (1–3), 156–181.

Sun, L., He, N., Yu, T., Duan, X., Feng, C. & Zhang, Y. 2017 The removal of typical pollutants in secondary effluent by the combined process of powdered activated carbon–ultrafiltration. *Water Science and Technology* **75** (6), 1485.

Tong, X., Cui, Y., Wang, Y. H., Bai, Y., Yu, T., Zhao, X. H., Ikuno, N., Luo, H. J., Hu, H. Y. & Wu, Y. H. 2020 Fouling properties of reverse osmosis membranes along the feed channel in an industrial-scale system for wastewater reclamation. *Science of the Total Environment* **713**, 9.

Wang, D., Hu, Q.-y., Li, M., Wang, C. & Ji, M. 2016 Evaluating the removal of organic fraction of commingled chemical industrial wastewater by activated sludge process augmented with powdered activated carbon. *Arabian Journal of Chemistry* **9**, S1951–S1961.

Wang, Y.-H., Wu, Y.-H., Tong, X., Yu, T., Peng, L., Bai, Y., Zhao, X.-H., Huo, Z.-Y., Ikuno, N. & Hu, H.-Y. 2019 Chlorine disinfection significantly aggravated the biofouling of reverse osmosis membrane used for municipal wastewater reclamation. *Water Research* **154**, 246–257.

Wang, X., Ma, J., Wu, Z. & Wang, Z. 2020 Stimulatory effects on bacteria induced by chemical cleaning cause severe biofouling of membranes. *Water Reuse* **10** (1), 82–94.

Xu, P., Bellona, C. & Drewes, J. E. 2010 Fouling of nanofiltration and reverse osmosis membranes during municipal wastewater reclamation: membrane autopsy results from pilot-scale investigations. *Journal of Membrane Science* **353** (1), 111–121.

Yang, T., Qiao, B., Li, G. C. & Yang, Q. Y. 2015 Improving performance of dynamic membrane assisted by electrocoagulation for treatment of oily wastewater: effect of electrolytic conditions. *Desalination* **363**, 134–143.

Yu, T., Xu, C., Chen, F., Yin, H. & Bi, X. 2021 Microcoagulation improved the performance of the UF–RO system treating the effluent from a coastal municipal wastewater treatment plant: a pilot-scale study. *Water Reuse* **11** (2), 177–188.

Zhang, M., Li, C., Benjamin, M. M. & Chang, Y. 2003 Fouling and natural organic matter removal in adsorbent/membrane systems for drinking water treatment. *Environmental Science & Technology* **37** (8), 1663.

Zhang, S., Yang, Y., Takizawa, S. & Hou, L.-a. 2018 Removal of dissolved organic matter and control of membrane fouling by a hybrid ferrihydrite-ultrafiltration membrane system. *Science of the Total Environment* **631–632**, 560–569.

Zhao, X., Hu, H.-Y., Yu, T., Su, C., Jiang, H. & Liu, S. 2014 Effect of different molecular weight organic components on the increase of microbial growth potential of secondary effluent by ozonation. *Journal of Environmental Sciences* **26** (11), 2190–2197.

Zhao, Y., Lu, D., Cao, Y., Luo, S., Zhao, Q., Yang, M., Xu, C. & Ma, J. 2018 Interaction analysis between gravity-driven ceramic membrane and smaller organic matter: implications for retention and fouling mechanism in ultralow pressure-driven filtration system. *Environmental Science & Technology* **52** (23), 13718–13727.

Zularisam, A. W., Ismail, A. F., Salim, M. R., Sakinah, M. & Ozaki, H. 2007 The effects of natural organic matter (NOM) fractions on fouling characteristics and flux recovery of ultrafiltration membranes. *Desalination* **212** (1), 191–208.

First received 19 July 2021; accepted in revised form 29 September 2021. Available online 20 October 2021

Water Recycling and Low-Carbon Development doi: 10.2166/wrd.2021.051

Effects of micro-flocculation pretreatment on the ultrafiltration membrane fouling caused by different dissolved organic matters in treated wastewater

Lanxin Ren[a], Chen Liu [a], Ting Meng [a] and Yingxue Sun [a,b,c,*]

[a] School of Ecology and Environment, Beijing Technology and Business University, Beijing 100048, China
[b] State Environmental Protection Key Laboratory of Food Chain Pollution Control, Beijing Technology and Business University, Beijing 100048, China
[c] Key Laboratory of Cleaner Production and Integrated Resource Utilization of China National Light Industry, Beijing Technology and Business University, Beijing 100048, China
*Corresponding author. E-mail: sunyx@th.btbu.edu.cn

CL, 0000-0002-0417-5073; TM, 0000-0001-6113-1037; YS, 0000-0002-3675-0706

ABSTRACT

This study investigated the efficacy of using micro-flocculation as a pretreatment approach in alleviating ultrafiltration (UF) membrane fouling caused by organic matter in treated wastewater. Three typical model dissolved organic matters (DOM), humic acid, fulvic acid, and sodium alginate, were employed to simulate membrane fouling. The results showed that micro-flocculation using poly aluminum chloride (PAC) or polymerized ferric sulfate (PFS) as flocculant could effectively enhance the treatment performance of the UF process on DOM. With 6 mg/L PAC, the removal efficiency of humic acid, fulvic acid, and sodium alginate by micro-flocculation combined UF process reached 79.95%, 63.25%, and 51.14%, respectively. Specifically, after micro-flocculation, micromolecular hydrophilic organic matter (e.g., fulvic acid) tended to form a compact cake layer. The macromolecular hydrophobic organic matter (e.g., humic acid) and macromolecular hydrophilic organic matter (e.g., sodium alginate) generally led to a loose cake layer. At PAC dosage of 6 mg/L, the membrane specific flux (J/J_0) at the end was improved by 11.71%, 10.27%, and 2.2% for humic acid, sodium alginate and fulvic acid solutions, respectively, compared with the UF process alone. It could be inferred that micro-flocculation pretreatment can effectively mitigate the membrane fouling when treating wastewater containing humic acid, sodium alginate, or fulvic acid.

Key words: dissolved organic matter (DOM), floc size, micro-flocculation, organic fouling, ultrafiltration (UF)

HIGHLIGHTS

- Micro-flocculation could enhance the capability of the UF process in removing DOM.
- Micro-flocculation pretreatment effectively alleviated UF membrane fouling.
- Cake layer played an important role in mitigating UF membrane fouling.
- The compactness of cake layer was related to the molecular weight, hydrophilicity or hydrophobicity of organic matters.

1. INTRODUCTION

Ultrafiltration (UF) is a reliable and promising technology owing to its guaranteed removal of suspended particles, colloids, and part of macromolecules for water treatment (Du *et al.* 2019; Takeuchi & Tanaka 2020). However, long-term operation of UF will cause membrane fouling (Ahmad *et al.* 2020), which is induced commonly by formation of a cake layer, concentration polarization, and membrane pore blocking (Wang *et al.* 2018a, 2018b; Bu *et al.* 2019). The UF membrane fouling is attributed to organic fouling and inorganic fouling (Wang & Tang 2011). Compared to inorganic membrane fouling, which is reversible and can be eliminated through permeate backwashing, the organic fouling resulting from dissolved organic matter (DOM) is a fatal inducement to UF membrane, which is tightly bounded to the membrane as irreversible fouling (Schulz *et al.* 2016; Wu *et al.* 2021).

The characteristics of DOM include hydrophobicity, molecular size, charge and solution chemistry, etc. which are likely to influence the formation of UF membrane fouling (Michael-Kordatou *et al.* 2015). However, there is a lack of consensus on the contribution of DOM fractions towards UF membrane fouling. Liu *et al.* (2014) found that the foulants larger than 1.2 μm had little effect on flux decline, but the molecular weight fraction between 100 kDa and 0.45 μm caused the most serious fouling. Some studies demonstrated that stronger hydrophobic parts were more responsible for UF membrane fouling than hydrophilic parts, but others reported that hydrophilic substances had a higher fouling potential (Shon *et al.* 2006; Tang *et al.* 2010). Tian *et al.* (2018) reported that reducing the proportion of macromolecular compound and improving the hydrophilic fractions can be effective approaches to mitigate membrane fouling.

A variety of pre-treatment technologies have been proposed and investigated to overcome the fouling issues during the UF treatment process, including coagulation (Bu *et al.* 2019), adsorption, and pre-oxidation (Cheng *et al.* 2016; Yu *et al.* 2018). Due to low cost, efficient colloidal removal, and reliable operation, the coagulation pretreatment is a widely used process (Yu *et al.* 2019). However, the coagulation process often involves a large number of coagulant dosages, a stirring process and a sedimentation process. Specifically, micro-flocculation is a kind of on-line coagulation pretreatment, with few dosages of flocculants, formation of micro-flocs (adsorbed to organic matter and retained by membrane) and without sedimentation, which can be directly combined with UF technology to remove organic matter and improve the membrane flux (Wang *et al.* 2018a, 2018b; Lee *et al.* 2021). In addition, micro-flocculation combined with UF process also has the advantages of removing low molecular weight organic matter and trace organic matter (Wang *et al.* 2018a, 2018b; Li *et al.* 2020). However, the difference of floc properties can lead to different degrees of membrane fouling (Feng *et al.* 2015). Also, the type of flocculants and operational parameters could affect the floc properties (Du *et al.* 2019).

Consequently, this paper aims to investigate the effects of micro-flocculation pretreatment on the fouling of UF membrane induced by different features of DOM. Three model DOM fractions, namely, humic acid (i.e., macromolecular and hydrophobic component), fulvic acid (i.e., micromolecular and hydrophilic component), and sodium alginate (i.e., macromolecular and hydrophilic component) were used to simulate the different influent quality of the micro-flocculation combined UF process. The looseness or closeness of the flocs have been analyzed in terms of different particle size distribution. Afterwards, UF membrane fouling behavior was evaluated through the measurement of the membrane permeate flux.

2. MATERIALS AND METHODS

2.1. Water samples

Three model DOM in water samples were simulated. These were humic acid (81.7%, Sinopharm Chemical Reagent Co., Ltd), fulvic acid (85%, Shanghai Aladdin Biochemical Technology Co., Ltd), and sodium alginate (Shanghai yuanye Bio-Technology Co., Ltd). Humic acid and fulvic acid stock solutions were prepared by 0.1 mol/L NaOH solution and then filtered with 0.22 μm membrane to remove insoluble ash. Sodium alginate stock solution (1 g/L) was prepared by ultra-pure water. The stock solutions were diluted as needed. The turbidity in solution was regulated by Kaolin (CNBM Geological Engineering Exploration Academy Co., Ltd), controlled in 10.0 ± 0.5 NTU. The pH was adjusted to 7.0 ± 0.2 by 0.1 mol/L HCl or NaOH. All the water used in the experiment was ultra-pure water. The water samples were stored at 4 °C in dark conditions.

2.2. Micro-flocculation test

Polymerized ferric sulfate (PFS) and poly aluminum chloride (PAC) (≥98%, Aladdin Industrial Corporation, Shanghai, China) were used as the flocculants. The micro-flocculation experiment was conducted in a ZR4-6 coagulation experimental mixer (Zhongrun Water Industry Technology Development Co., Shenzhen, China) at 25 °C. The experimental procedure was as follows: mixing at fast speed of 200 r/min for 30 s after adding the flocculant (PAC or PFS), and mixing at slow speed of 40 r/min for 15 min. Afterwards, the experiments were conducted using two approaches: (1) immediately taking water samples for UF process (shown in Section 2.3) or floc size test; (2) settling for 30 min and taking the appropriate volume of water samples to measure the dissolved organic carbon (DOC).

2.3. UF process test

The UF process was simulated using a cup ultrafilter (Amicon 8400, Millipore, USA) with a capacity of 400 mL. The pressure was provided by nitrogen and was maintained at 0.10 MPa. A circular regenerated cellulose membrane (Millipore, plgc07610, USA) was used. The diameter was 76 mm, the effective membrane area was 41.8 cm^2, and the molecular weight cutoff was 10 kDa. The electronic balance was connected to the computer, and the mass of accumulated filtrated water was recorded in real time at intervals of 60 s. The membrane flux (J) was calculated by cumulative water output. The new membrane was immersed in ultra-pure water for at least 24 h, in which the water was changed every 8 h to remove the protective substances on the membrane surface.

The ultra-pure water was filtrated under 0.10 MPa until the effluent flux was stable, which was the initial flux of membrane (J_0), then the water samples after micro-flocculation were filtrated under the same conditions, which was the permeate flux of membrane (J). The normalized specific flux J/J_0 was used to represent the membrane performance as a function of time. After the UF experiment, the permeate was collected and the DOC, turbidity and pH in permeate were further measured.

2.4. Analysis methods

Turbidity and pH were measured using a 2100Q turbid meter and PHC101 pH meter (Hash, USA), respectively. DOC was measured using a total organic carbon analyzer (Vario-TOC, GER). The Zeta potential of the solution was tested by nano sizer and Zeta-potential tester (Malvern Zetasizer Nano ZS90, UK).

A laser diffraction instrument SALD-2300 (Shimadzu, JPN) was used to monitor the change of floc size. Floc fractal dimension (D_f) was calculated by the particle size distribution method, and the function relationship between cumulative distribution function $p(r)$ and particle size d_p was calculated by Equation (1):

$$p(r) = \frac{N(r)}{N} = \frac{k}{N(1-\beta)} \cdot r^{1-\beta} \tag{1}$$

The natural logarithm was calculated by Equation (2):

$$\ln[p(r)] = D_f \cdot \ln r + \ln K \tag{2}$$

where $N(r)$ is number of flocs with particle size less than or equal to r; N is total number of floc particles in water; r is particle size; k, β, and K are constant.

The permeate flux of the membrane (J) can clearly be presented as shown in Equation (3):

$$J = \frac{Q}{A_t} = \frac{M}{\rho} \cdot \frac{1}{A_t} \tag{3}$$

where J ($L/m^2{\cdot}h$) is the permeate flux of the membrane; Q (L) is liquid permeation; A (m^2) is the effective membrane area; t (h) is the time; M (g) is the quality of outflow water for specific time; ρ (kg/m^3) is the density of outflow water.

The surface morphology of UF membrane was observed with a scanning electron microscope (SEM, ZEISS Gemini 300, GER). Before SEM observation, samples were dried in a desiccator (45 °C) for 72 h. The surface roughness and morphology of floc deposition layer were measured by atomic force microscope (AFM, Olympus LEXT OLS4500 JPK NW4, JPN). The hydrophilicity of the fouled membranes was measured by a dynamic contact angle tensiometer (Biolin Theta Flex, SWE).

3. RESULTS AND DISCUSSION

3.1. Performance of the micro-flocculation combined UF process

The effects of different flocculants of PAC and PFS in the micro-flocculation combined UF process on the removal of organic matter in water samples are shown in Figure 1(a) and 1(b), respectively. The changes of pH in the micro-flocculation combined UF process with different flocculants were evaluated, as shown in Figure 2. The removal rates of humic acid, fulvic acid, and sodium alginate were 83.52%, 56.67%, and 43.66%, respectively, with direct UF process. As for PAC-based micro-flocculation combined UF process, the highest DOC removal efficiency was found at PAC dosage of 1 mg/L, and the removal efficiency on humic acid, sodium alginate, and fulvic acid was 86.39%, 86.75%, and 64.90%, respectively. However, with the further increase of PAC dosage (2–6 mg/L), the removal efficiency of the three model DOM was reduced and the pH of the solution had a gentle decrease after the reaction, which could be explained by the charge neutralization between negatively charged natural organic matter and positively charged aluminum hydrolysates (Wang *et al.* 2020). Particularly, at PAC dosage of 6 mg/L, the removal efficiency of humic acid, sodium alginate, and fulvic acid was 79.95%, 63.25%, and 51.14%, respectively. Deng *et al.* (2019) discovered that pretreatment is ineffective in eliminating micromolecular or hydrophilic components at the PAC dosage of 10 mg/L. Likewise, the addition of PFS as flocculant presents a similar tendency on the removal of the three model DOM. At the PFS dosage of 1 mg/L, the DOC removal efficiency of humic acid, sodium alginate, and fulvic acid was 85.55%, 70.18%, and 63.43%, respectively, while at the PFS dosage of 6 mg/L, the

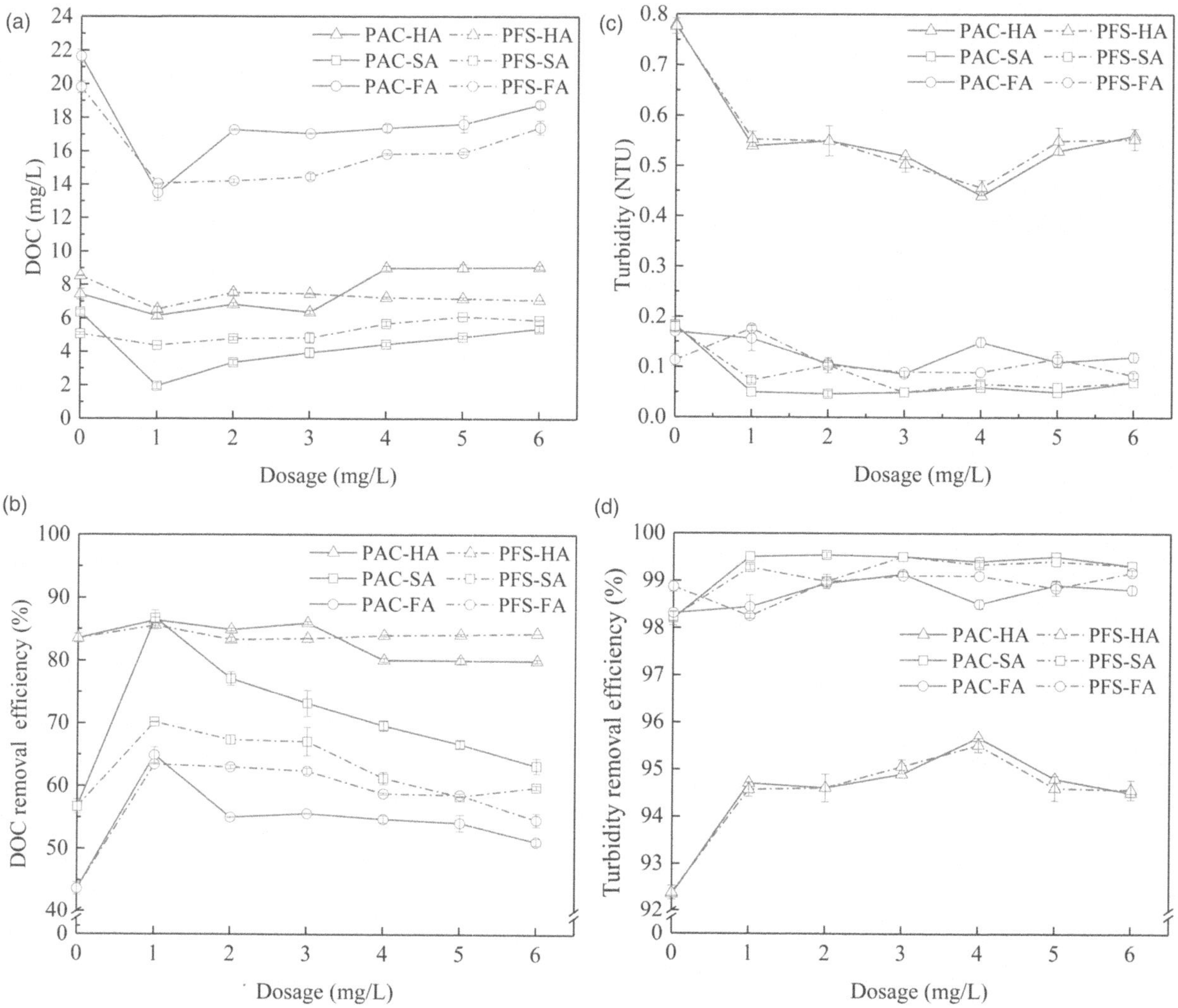

Figure 1 | Effects of flocculants in the micro-flocculation combined UF process on DOC removal in the effluent (a) and DOC removal efficiency (b), turbidity (c), and turbidity removal efficiency (d). Condition: the concentration of NOM (calculated by DOC), [HA] = 45.17 mg/L (HA represented humic acid), [FA] = 38.42 mg/L (FA represented fulvic acid), [SA] = 14.68 mg/L (SA represented sodium alginate), initial pH in PAC and PFS experiment at 7.05 and 7.03, respectively, initial turbidity in HA and SA experiment at 10.2 NTU, FA experiment at 10.1 NTU, temperature = 25 °C, running time of UF = 100 min, J_0 = 51.75 (L/m^2·h).

removal efficiency of the humic acid, sodium alginate, and fulvic acid was 84.31%, 59.82%, and 54.66%, respectively. After adding PFS as flocculant, the pH of the solution also showed a trend, similar to the effect of PAC.

Figure 1(c) and 1(d) show the changes of the effluent turbidity after the micro-flocculation combined UF process. With the increase of the flocculant dosage, the turbidity removal tended to be stable. Compared with the direct UF process, micro-flocculation prior to UF contributed slightly to the improvement on removal of turbidity of the effluent. The effects of PFS and PAC on turbidity removal were similar. Specifically, the effluent turbidity of water samples was lower than 0.8 NTU. Nevertheless, when the flocculant was added continuously, negative impacts on water turbidity may occur, since the flocs with large and the same charge were difficult to settle (Shen *et al.* 2020).

In summary, when the flocculants were added, the UF process could have a better removal efficiency on the three model DOM, which indicated that the micro-flocculation facilitated the removal of organic pollutants and hydrophobic ones were more easily removed by micro-flocculation UF than hydrophilic ones. However, the higher flocculant dose could not further improve removal efficiency. Moreover, the regenerated cellulose membrane is a hydrophilic membrane which is beneficial to the DOC removal in terms of hydrophobic organic matter (Yu *et al.* 2018).

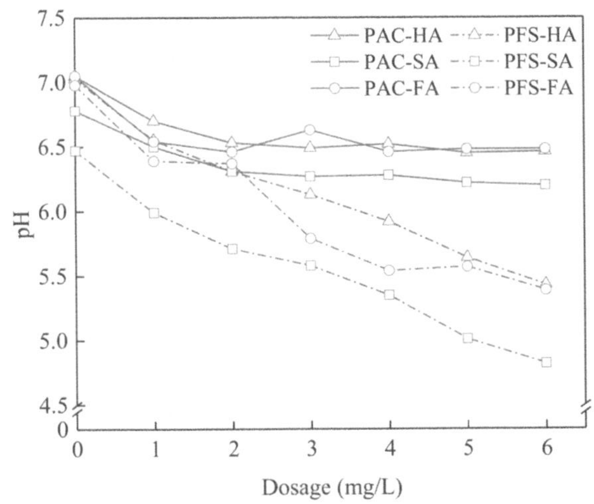

Figure 2 | The changes of pH in the different flocculants' micro-flocculation combined UF process. Condition: the concentration of DOM (calculated by DOC), [HA] = 45.17 mg/L, [FA] = 38.42 mg/L, [SA] = 14.68 mg/L, initial pH in PAC and PFS experiment at 7.05 and 7.03, respectively, initial turbidity in HA and SA experiment at 10.2 NTU, FA experiment at 10.1 NTU, temperature = 25 °C, running time of UF = 100 min, J_O = 51.75 (L/m^2·h).

In this study, it was found that the PAC with a dosage of 6 mg/L had the worst removal effects on three model DOM with the change of flocculants and its dosages, where there may have serious membrane fouling. Thus, PAC with a dosage of 6 mg/L was chosen to further study the influence of micro-flocculation on the pollution of the UF membrane.

3.2. Effect of micro-flocculation on the UF membrane flux

The influence of the flocs on membrane filtration behavior, membrane specific flux of suspensions with filtration time under constant pressure is shown in Figure 3.

The membrane-specific flux for humic acid is shown in Figure 3(a), in which a gentle decrease from 0.8405 to 0.8126 was observed within 540 min UF cycle and the attenuation rate was 3.32%, which suggested that the humic acid was more likely to block the membrane pores at the beginning of UF. When the dosage of PAC was 6 mg/L, initial and ending membrane-specific flux was 0.9466 and 0.9204, respectively, and the attenuation rate was 2.8%, showing a gentle decrease. In addition, the membrane-specific flux at the end was improved by 11.71% compared with UF alone, which indicated that the micro-flocculation had a positive effect on alleviating the membrane fouling. This may be caused by the cake layer produced by the combination of flocculant and the organic matters, then sedimentation on the surface of the UF membrane effectively prevented direct contact between the pollutants and the UF membrane (Wang *et al.* 2019; Wang *et al.* 2020), which slowed down the fouling caused by the absorption and blockage of membrane pores (Yu *et al.* 2018).

Figure 3(b) presents the membrane-specific flux for fulvic acid after adding PAC as the flocculant. The flux attenuated from 0.9736 to 0.8766 and the attenuation rate was 9.96%. Compared with the UF alone, the initial membrane-specific flux increased by 0.0721 but the attenuation rate increased by 5.05%. Although the membrane-specific flux at the end was improved by 2.2% compared with UF alone, the degree of membrane fouling was exacerbated and the PAC may have a negative impact on mitigating the membrane fouling in fulvic acid solution. That could be explained by the fact that, owing to the compression of the electric double layer, the negatively charged fulvic acid becomes coiled (Shen *et al.* 2020), which induced formation of a cake layer with a dense structure, which could be confirmed from Table 1. Fulvic acid has the largest floc fractal dimension. In addition, the formed small flocs produced by micromolecular fulvic acid are likely to block the membrane pore or the pores of the cake layer, resulting in a faster rate of membrane fouling.

As shown in Figure 3(c), the membrane-specific flux of UF decayed from 0.8682 to 0.7656, where the attenuation rate was 11.82%. When the dosage was 6 mg/L, the membrane-specific flux attenuated from 0.9570 to 0.8532, and its attenuation rate was 10.85%. The membrane-specific flux at the end was improved by 10.3% compared with UF alone. This indicated that adding PAC could reduce membrane fouling that was the result of sodium alginate. This is related to the loose flocs formed by fulvic acid (Xiong *et al.* 2019), as shown in Table 1. However, the membrane-specific flux at the end was still at

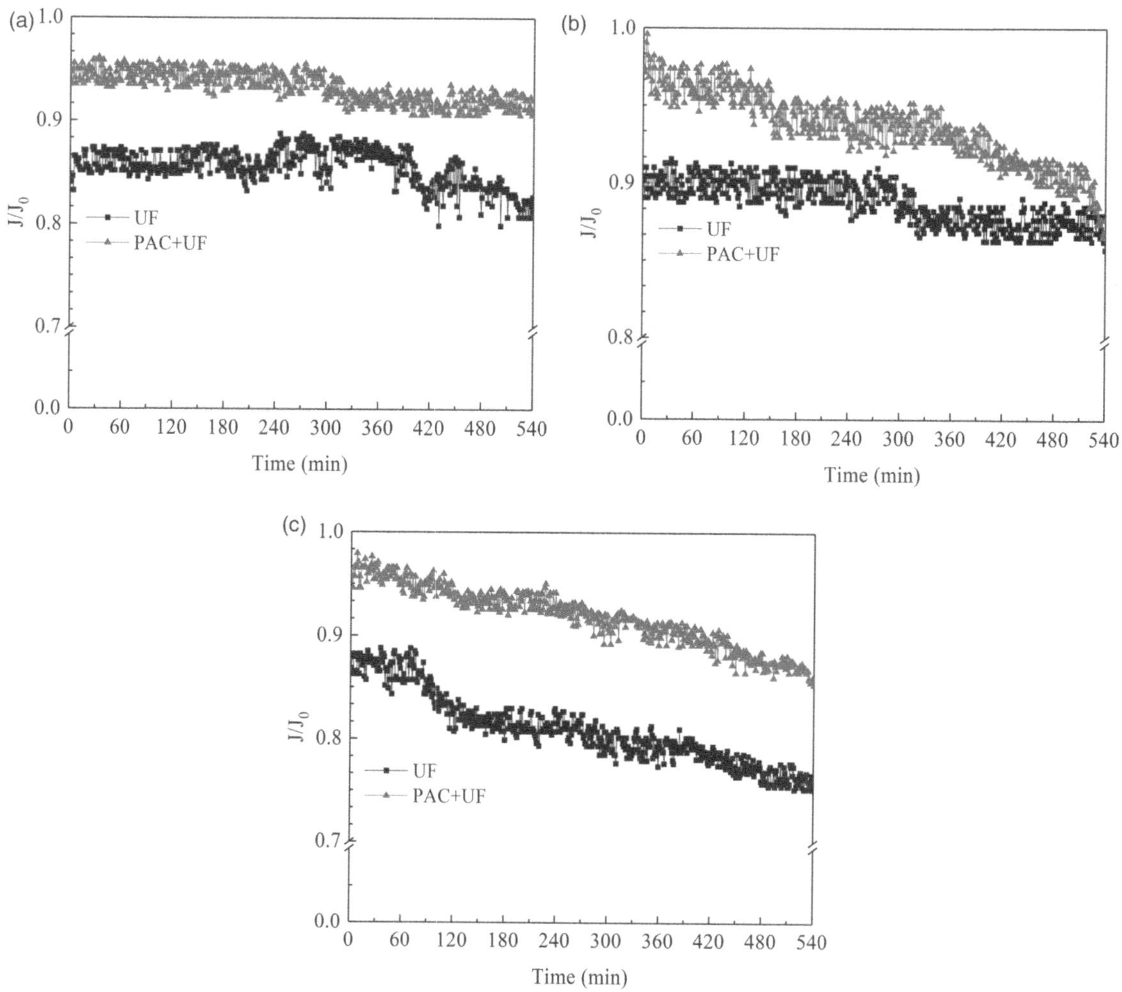

Figure 3 | The decay curve of membrane-specific flux by UF membrane: HA (a), FA (b), and SA (c). Condition: PAC = 6 mg/L, the concentration of DOM (calculated by DOC), [HA] = 45.17 mg/L, [FA] = 38.42 mg/L, [SA] = 14.68 mg/L, initial pH in PAC experiment at 7.05, initial turbidity in HA and SA experiment at 10.2 NTU, FA experiment at 10.1 NTU, temperature = 25 °C, J_O = 51.75 (L/m^2·h).

Table 1 | Floc fractal dimension of the PAC on different DOM

Dissolved organic matter	Floc fractal dimension
Humic acid	0.8786
Fulvic acid	0.9765
Sodium alginate	0.8163

a relatively low level. This may be because the uniform gel-like substance with low permeability formed by the terminal carboxyl of sodium alginate preferentially formed coordination effect with metal ions. In addition, as a hydrophilic organic matter, it could be difficult to remove by a hydrophilic membrane, which was similar to fulvic acid.

For the three DOM, raw water without pretreatment may cause serious membrane fouling. The micro-flocculation as a pretreatment approach prior to UF membrane could, indeed, alleviate the membrane fouling issue, and the removal effects on macromolecular components (e.g., humic acid and sodium alginate) were better than those of micromolecular components (e.g., fulvic acid). Notably, the hydrophilic organics (e.g., fulvic acid and sodium alginate) were more likely to cause membrane fouling on hydrophilic membranes than hydrophobic organics (e.g., humic acid).

3.3. The floc properties on the UF membrane fouling

Some studies found that most flocs could accumulate on the membrane surface and lead to the formation of cake layer, which was beneficial to the removal of contaminants (Wang *et al.* 2019). However, flocs close to the membrane pores may cause serious membrane fouling due to pore clogging (Yu *et al.* 2017). Flocs that are significantly larger than the membrane pore size are unlikely to cause significant membrane fouling (Guo *et al.* 2012). Therefore, the floc size distribution and the floc fractal dimension after micro-flocculation were analyzed to indirectly reflect the nature of membrane fouling.

The analytical results are shown in Figure 4. After micro-flocculation, the floc size distribution formed by humic acid, fulvic acid, and sodium alginate was 0–32 μm (Figure 4(a)), 0–12 μm (Figure 4(b)), and 0–40 μm (Figure 4(c)), respectively. The average particle sizes of the formed flocs were sodium alginate (7.54 μm), humic acid (4.99 μm), and fulvic acid (3.12 μm) in descending order. The floc fractal dimension represents the degree of looseness and compactness (Yao *et al.* 2015; Wang *et al.* 2018a, 2018b). Also, the smaller the floc fractal dimension, the looser the flocs formed. As shown in Table 1, the floc fractal dimension produced by humic acid, fulvic acid and sodium alginate was 0.8786, 0.9765, and 0.8163, respectively, which showed that the flocs formed by fulvic acid were more compact than humic acid and sodium alginate.

This could be explained by the fact that owing to the relatively loose cake layer produced by humic acid compared with fulvic acid, the binding sites were available on the humic acid molecules and on the surface of the PAC (Wang *et al.* 2018a, 2018b), which caused the adsorption of humic acid in the cake layer (Wang *et al.* 2019) and the high DOC removal efficiency of humic acid (Figure 1(b)). Cho *et al.* (2006) discovered that small flocs could aggravate membrane fouling. The compact flocs formed by fulvic acid could cause the cake layer to become less porous and highly resistant on the surface of UF membrane (Shen *et al.* 2020). This could explain the phenomenon that the tight cake layer was not conducive to DOC removal. According to previous research, the formation of loose and multi-branched flocs helps to form a porous and

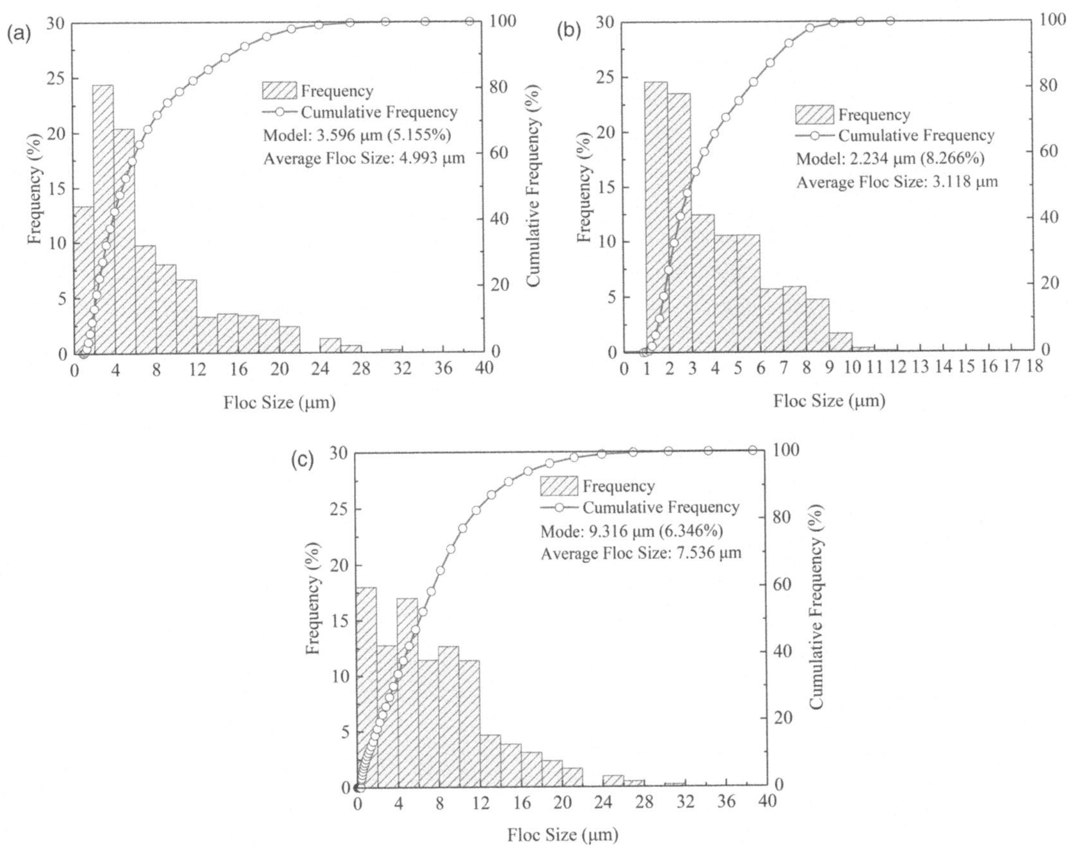

Figure 4 | Floc particle size distribution and cumulative frequency with adding 6 mg/L PAC: HA (a), FA (b), and SA (c). Condition: the concentration of DOM (calculated by DOC), [HA] = 45.17 mg/L, [FA] = 38.42 mg/L, [SA] = 14.68 mg/L, initial pH in PFS experiment at 7.03, initial turbidity in HA and SA experiment at 10.2 NTU, FA experiment at 10.1 NTU, temperature = 25 °C, J_0 = 51.75 (L/m^2·h), running time of UF = 540 min.

fluffy cake layer and alleviate membrane fouling produced by sodium alginate (Roth *et al.* 2015). This conclusion was consistent with the phenomenon shown in Figure 3(c).

In summary, the average particle size of the flocs formed by the macromolecular component was significantly larger than that of the micromolecular component. Moreover, the micromolecular components were more inclined to form compact flocs, while macromolecular components were more likely to form loose flocs.

The zeta potential of the flocs was examined to evaluate the ability of the flocculant to destabilize the particles (Dayarathne *et al.* 2020). According to Figure 5, within the selected PAC dosage (1–6 mg/L), the zeta potentials of the three model DOM were all negative, and reached the lowest at the PAC dosage of 6 mg/L. It was observed that the zeta potentials in humic acid, fulvic acid, and sodium alginate samples varied from −47.4 mV to −44.7 mV, −28.7 mV to −26.2 mV, and −61.0 mV to −47.9 mV, respectively. The decrease in negative zeta potential indicates a decrease in repulsive electrostatic forces (Chekli *et al.* 2017). This indicated that charge neutralization may be the main mechanism in the flocculation of DOM by PAC (Wang *et al.* 2019; Shen *et al.* 2020).

3.4. Characterizations of the cake layer on UF membranes

3.4.1. Membrane surface morphology

The surface topographies of pristine and fouled membranes were characterized using SEM, and the images are shown in Figure 6. It could be observed that the pristine membrane exhibited a clean and smooth surface without any foulants accumulated (Figure 6(a)). After the direct UF process of different DOM solutions, the membrane surface morphologies became different due to the accumulation of the foulants. However, as can be seen from Figure 6(b) and 6(f), the membrane fouled by humic acid and sodium alginate solutions without pretreatment did not show obvious variations. Comparatively, as shown in Figure 6(d), a dense and compact fouling layer was present on the membrane surface with fulvic acid solution. According to the results of membrane-specific flux of humic acid and sodium alginate, the formation of internal fouling in membrane was inferred (Wang *et al.* 2019; Wan *et al.* 2021).

After the micro-flocculation, humic acid was deposited and a relatively dense cake layer was formed on the membrane surface (Wu *et al.* 2020), where the obvious cracks appeared after the drying procedure, as shown in Figure 6(c). However, the membrane surface of the pretreated fulvic acid and sodium alginate showed a smoother, dense, and without obvious porous structure (Xiong *et al.* 2019; Long *et al.* 2021).

Combining the experimental results of floc fractal dimension and membrane flux, it could be inferred that sodium alginate forms a more permeable and loose cake layer on the membrane surface, which could effectively reduce membrane fouling (Zhao *et al.* 2020). In addition, the humic acid showed similar results. Although the SEM images of fulvic acid and sodium alginate were relatively similar, the cake layer produced by fulvic acid was more compact and uniform. This may

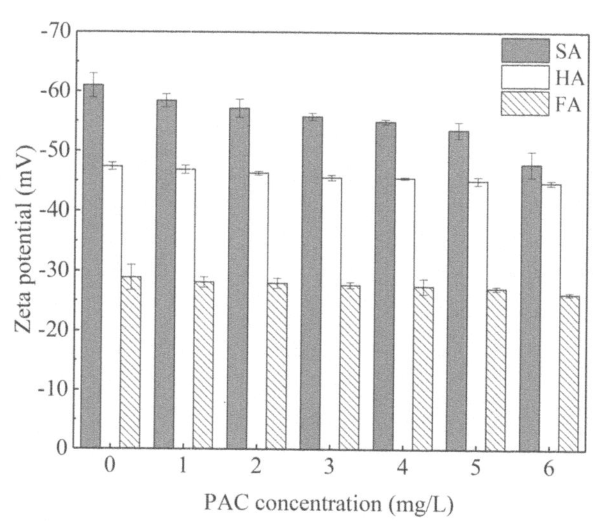

Figure 5 | Zeta potentials of PAC (a) and PFS (b) in the micro-flocculation combined UF process. Condition: the concentration of DOM (calculated by DOC), [HA] = 45.17 mg/L, [FA] = 38.42 mg/L, [SA] = 14.68 mg/L, initial pH in PAC and PFS experiment at 7.05 and 7.03, respectively, initial turbidity in HA and SA experiment at 10.2 NTU, FA experiment at 10.1 NTU, temperature = 25 °C, J_0 = 51.75 (L/m^2·h).

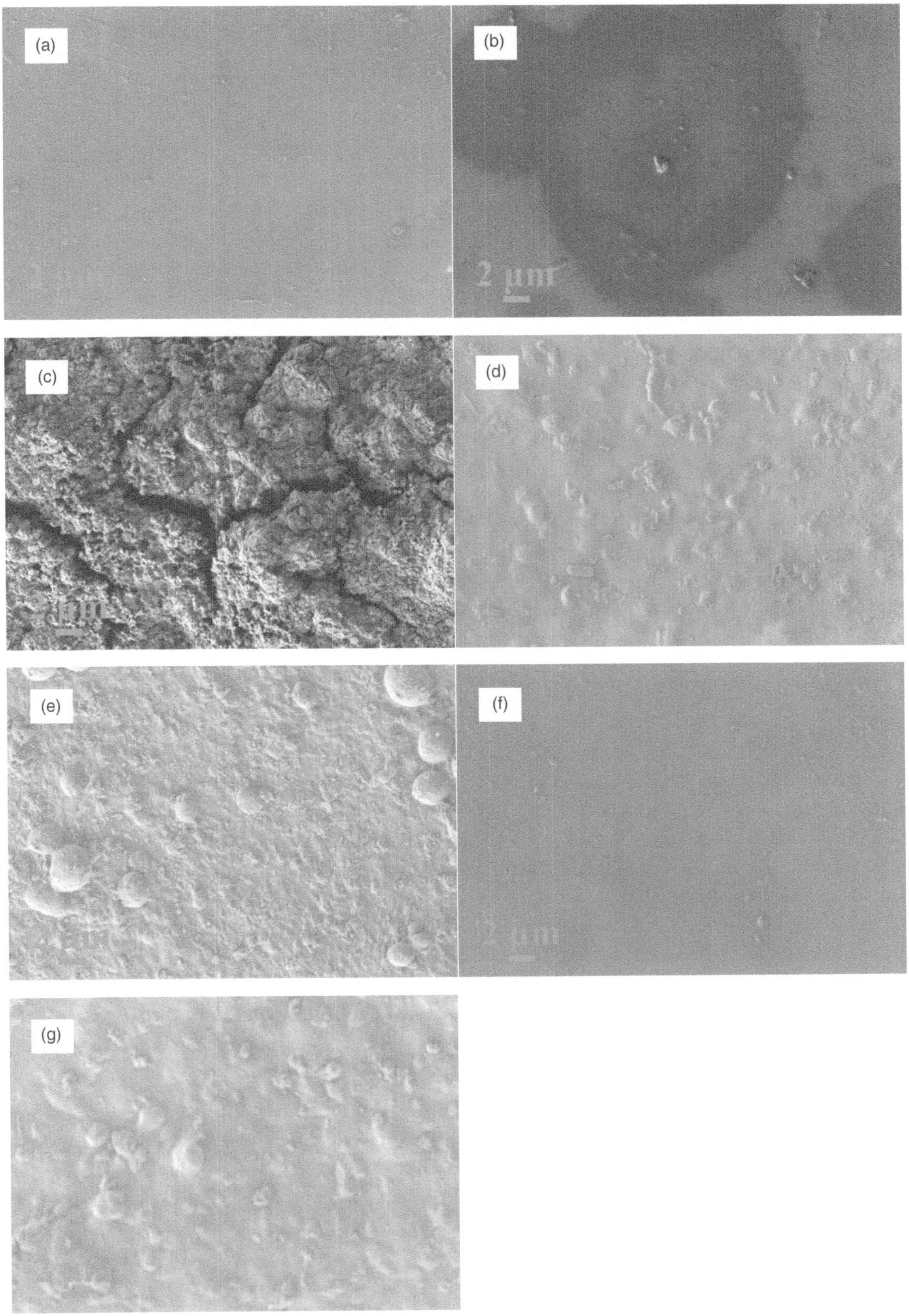

Figure 6 | Morphology of membrane surface: pristine membrane (a), filtering HA raw water (b), filtering HA raw water added with 6 mg/L PAC (c), filtering FA raw water (d), filtering FA raw water added with 6 mg/L PAC (e), filtering SA raw water (f), filtering SA raw water added with 6 mg/L PAC (g). Condition: the concentration of NOM (calculated by DOC), [HA] = 45.17 mg/L, [FA] = 38.42 mg/L, [SA] = 14.68 mg/L, initial pH in PAC and PFS experiment at 7.05 and 7.03, respectively, initial turbidity in HA and SA experiment at 10.2 NTU, FA experiment at 10.1 NTU, temperature = 25 °C, J_0 = 51.75 (L/m^2·h), running time of UF = 540 min.

be explained by the fact that fulvic acid, as micromolecular organic matter, easily led to the small size of the formed flocs. Combined with the results shown in Figure 3(b) that fulvic acid solution was associated with a larger membrane flux attenuation rate, it could be inferred that some small flocs blocked the membrane pores and hindered the removal of fulvic acid.

3.4.2. Contact angle of fouled membranes on different DOM

The contact angle is used to evaluate the hydrophilicity and hydrophobicity of the material. The small membrane contact angle corresponds to a high hydrophilicity (Thomas *et al.* 2014).

As can be seen in Table 2, the contact angle of the pristine membrane was 30.4°, indicating that regenerated cellulose was hydrophilic membrane (Tlili & Alkanhal 2019). After filtering the humic acid, fulvic acid, and sodium alginate solution, the contact angle of fouling membrane increased apparently to 69.9°, 61.5°, and 41.2°, respectively (Figure S1 in the supplementary material and Table 2). This happened because the surface of the membrane became more hydrophobic due to the formation of the fouling on UF membrane. According to previous studies, the hydrophobicity of membrane surface is closely related to the presence of aromatic compound, and high aromatic compound content could enhance the hydrophobicity (Mu *et al.* 2019). However, compared to the UF process alone, after employing micro-flocculation as pretreatment, the contact angles became smaller, with 57.6°, 39.3°, and 40.2° in humic acid, fulvic acid, and sodium alginate solutions, respectively. It can be inferred that the micro-flocculation pretreatment improved the hydrophilicity of cake layer. Some studies have suggested that the formation of the hydrophilic cake layer is beneficial to reduce membrane fouling (Sun *et al.* 2018).

3.4.3. Surface roughness and morphology of floc deposition layer

The roughness of surface is an important factor affecting the membrane fouling resistance, which is represented by Ra (Yu *et al.* 2019). The surface morphology of the pristine membrane and the contaminated film were analyzed by AFM (Li *et al.* 2021), as shown in Figure 7.

The convex structures on the surface of pristine membrane (Figure 7(a)) were uniformly distributed, with an Ra of 3.3. After UF with humic acid alone, the surface roughness of the membrane increased. When PAC was added, the protrusions on the surface area increased (Figure 7(c)), with an Ra of 27.7. According to Figure 3, the steady-state membrane-specific flux of raw water was lower than that with the addition of PAC, which indicated that the cake layer became more porous. The SEM results also confirmed the findings. After filtrating fulvic acid directly, the surface roughness of the membrane increased significantly, with an Ra of 54.7, and a large amount of sediment adhered to the membrane surface. With the addition of PAC, it was observed that the cake layer became tighter (Figure 7(e)), the Ra decreased to 30.5, and the steady-state membrane had a larger attenuation rate. This is consistent with the results obtained in Table 2. Interestingly, after filtrating the sodium alginate solution directly (Figure 7(f)), the surface roughness of the membrane was smaller than pristine membrane and the Ra was 1.5. Combined with SEM findings, it could be concluded that the convex structure on the pristine membrane surface was smoother and denser. When treating sodium alginate solution, despite the fact that the membrane surface was covered and wrapped by contaminants, little variation on surface roughness was observed since the sodium alginate solution is viscous.

Basically, after adding PAC, Ra of three DOMs increased significantly, and the steady-state membrane-specific flux was higher than raw water. This implied that the membrane-specific flux was related to the roughness of the membrane surface.

4. CONCLUSIONS

This study focused on the removal efficiency of DOM and the change of membrane performance induced by different organic matters through micro-flocculation pretreatment. The micro-flocculation pretreatment was found to be effective to enhance the capability of the UF process in removing DOM. The hydrophilic membranes were able to achieve better removal efficiency on hydrophobic organic matters over hydrophilic organic matter. During the sole UF process, the membrane pores

Table 2 | Contact angle under different processing conditions

	Ultra-pure water	Humic acid	Fulvic acid	Sodium alginate
UF	30.4	69.9	61.5	41.2
PAC + UF		57.6	39.9	40.2

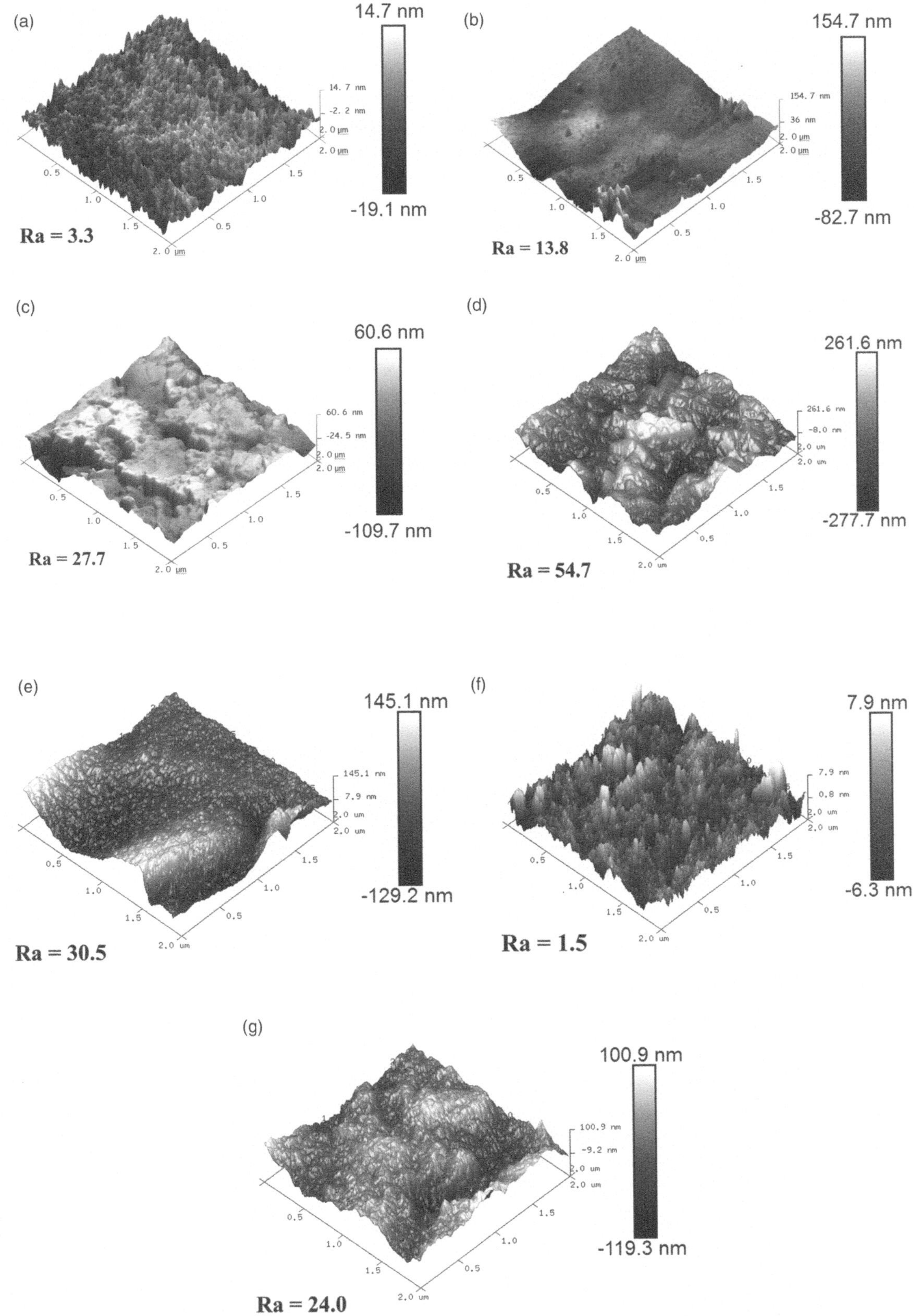

Figure 7 | The surface roughness and morphology of floc deposition layer: pristine membrane (a), filtering HA raw water (b), filtering HA raw water added with 6 mg/L PAC (c), filtering FA raw water (d), filtering FA raw water added with 6 mg/L PAC (e), filtering SA raw water (f), filtering SA raw water added with 6 mg/L PAC (g). Condition: the concentration of DOM (calculated by DOC), [HA] = 45.17 mg/L, initial pH in PAC experiment at 7.05, initial turbidity in HA experiment at 10.2 NTU, temperature = 25 °C, J_O = 51.75 (L/m^2·h), running time of UF = 540 min.

were easily clogged by DOM, but after applying micro-flocculation as the pretreatment approach, the increase of membrane-specific flux was observed, suggesting that the cake layer could alleviate membrane fouling. In addition, feed solution with micromolecular hydrophilic organic matter (e.g., fulvic acid) was more inclined to induce a compact cake layer. Noticeably, a relatively loose cake layer was formed by macromolecular hydrophobic organic matter (e.g., humic acid) or macromolecular hydrophilic organic matter (e.g., sodium alginate). The hydrophilic organics (e.g., fulvic acid and sodium alginate) were more likely to cause membrane fouling on hydrophilic membranes over hydrophobic organics (e.g., humic acid). In addition, the micro-flocculation could improve the hydrophilicity of the cake layer. Overall, incorporating micro-flocculation as a pre-treatment prior to UF plays an important role in alleviating the membrane fouling of the UF membrane and improving the removal of organic matter during the UF treatment process.

DATA AVAILABILITY STATEMENT

All relevant data are included in the paper or its Supplementary Information.

REFERENCES

Ahmad, M. A., Zainal, B. S., Jamadon, N. H., Yaw, T. C. S. & Abdullah, L. C. 2020 Filtration analysis and fouling mechanisms of PVDF membrane for POME treatment. *Journal of Water Reuse and Desalination* **10** (3), 187–199. doi:10.2166/wrd.2020.101.

Bu, F., Gao, B., Shen, X., Wang, W. & Yue, Q. 2019 The combination of coagulation and ozonation as a pre-treatment of ultrafiltration in water treatment. *Chemosphere* **231**, 349–356. doi:10.1016/j.chemosphere.2019.05.154.

Chekli, L., Corjon, E., Tabatabai, S. A. A., Naidu, G., Tamburic, B., Park, S. H. & Shon, H. K. 2017 Performance of titanium salts compared to conventional $fecl_3$ for the removal of algal organic matter (AOM) in synthetic seawater: coagulation performance, organic fraction removal and floc characteristics. *Journal of Environmental Management* **201**, 28–36. doi:10.1016/j.jenvman.2017.06.025.

Cheng, X., Liang, H., Ding, A., Qu, F., Shao, S., Liu, B., Wang, H., Wu, D. & Li, G. 2016 Effects of pre-ozonation on the ultrafiltration of different natural organic matter (NOM) fractions: membrane fouling mitigation, prediction and mechanism. *Journal of Membrane Science* **505**, 15–25. doi:0.1016/j.memsci.2016.01.022.

Cho, M., Lee, C. & Lee, S. 2006 Effect of flocculation conditions on membrane permeability in coagulation–microfiltration. *Desalination* **191** (1–3), 386–396. doi:10.1016/j.desal.2005.08.017.

Dayarathne, H. N. P., Angove, M. J., Aryal, R., Abuel-Naga, H. & Mainali, B. 2020 Removal of natural organic matter from source water: review on coagulants, dual coagulation, alternative coagulants, and mechanisms. *Journal of Water Process Engineering* **40**, 101820. doi:10.1016/j.jwpe.2020.101820.

Deng, L., Ngo, H., Guo, W. & Zhang, H. 2019 Pre-coagulation coupled with sponge-membrane filtration for organic matter removal and membrane fouling control during drinking water treatment. *Water Research* **157**, 155–166. doi:10.1016/j.watres.2019.03.052.

Du, P., Li, X., Yang, Y., Su, Z., Li, H., Wang, N., Guo, T., Zhang, T. & Zhou, Z. 2019 Optimized coagulation pretreatment alleviates ultrafiltration membrane fouling: the role of floc properties and slow-mixing speed on mechanisms of chitosan-assisted coagulation. *Journal of Environmental Sciences* **82**, 82–92. doi:10.1016/j.jes.2019.02.027.

Feng, L., Wang, W., Feng, R., Zhao, S., Dong, H., Sun, S., Gao, B. & Yue, Q. 2015 Coagulation performance and membrane fouling of different aluminum species during coagulation/ultrafiltration combined process. *Chemical Engineering Journal* **262**, 1161–1167. doi:10.1016/j.cej.2014.10.078.

Guo, W., Ngo, H. & Li, J. 2012 A mini-review on membrane fouling. *Bioresource Technology* **122**, 27–34. doi:10.1016/j.biortech.2012.04.089.

Lee, H., Lim, B., Hur, J., Kim, H. & Shin, H. 2021 Combined dual-size foam glass media filtration process with micro-flocculation for simultaneous removal of particulate and dissolved contaminants in urban road runoff. *Journal of Environmental Management* **277**, 111475. doi:10.1016/j.jenvman.2020.111475.

Li, R., Gao, B., Wang, W., Yue, Q. & Wang, Y. 2020 Floc properties and membrane fouling in coagulation/ultrafiltration process for the treatment of Xiaoqing River: the role of polymeric aluminum-polymer dual-coagulants. *Chemosphere* **243**, 125391. doi:10.1016/j.chemosphere.2019.125391.

Li, R., Li, J., Rao, L., Lin, H., Shen, L., Xu, Y., Chen, J. & Liao, B. 2021 Inkjet printing of dopamine followed by UV light irradiation to modify mussel-inspired PVDF membrane for efficient oil-water separation. *Journal of Membrane Science* **619**, 118790. doi:10.1016/j.memsci.2020.118790.

Liu, Y., Li, X., Yang, Y., Ye, W., Ji, S., Ren, J. & Zhou, Z. 2014 Analysis of the major particle-size based foulants responsible for ultrafiltration membrane fouling in polluted raw water. *Desalination* **347**, 191–198. doi:10.1016/j.desal.2014.05.039.

Long, Y., Yu, G., Dong, L., Xu, Y., Lin, H., Deng, Y., You, X., Yang, L. & Liao, B. 2021 Synergistic fouling behaviors and mechanisms of calcium ions and polyaluminum chloride associated with alginate solution in coagulation-ultrafiltration (UF) process. *Water Research* **189**, 116665. doi:10.1016/j.watres.2020.116665.

Michael-Kordatou, I., Michael, C., Duan, X., He, X., Dionysiou, D. D., Mills, M. A. & Fatta-Kassinos, D. 2015 Dissolved effluent organic matter: characteristics and potential implications in wastewater treatment and reuse applications. *Water Research* **77**, 213–248. doi:10.1016/j.watres.2015.03.011.

Mu, S., Wang, S., Liang, S., Xiao, K., Fan, H., Han, B., Liu, C., Wang, X. & Huang, X. 2019 Effect of the relative degree of foulant 'hydrophobicity' on membrane fouling. *Journal of Membrane Science* **570–571**, 1–8. doi:10.1016/j.memsci.2018.10.023.

Roth, E. J., Gilbert, B. & Mays, D. C. 2015 Colloid deposit morphology and clogging in porous media: fundamental insights through investigation of deposit fractal dimension. *Environmental Science & Technology* **49** (20), 12263–12270. doi:10.1021/acs.est.5b03212.

Schulz, M., Soltani, A., Zheng, X. & Ernst, M. 2016 Effect of inorganic colloidal water constituents on combined low-pressure membrane fouling with natural organic matter (NOM). *Journal of Membrane Science* **507**, 154–164. doi:10.1016/j.memsci.2016.02.008.

Shen, X., Gao, B., Guo, K. & Yue, Q. 2020 Characterization and influence of floc under different coagulation systems on ultrafiltration membrane fouling. *Chemosphere* **238**, 124659. doi:10.1016/j.chemosphere.2019.124659.

Shon, H. K., Vigneswaran, S., Kim, I. S., Cho, J. & Ngo, H. H. 2006 Fouling of ultrafiltration membrane by effluent organic matter: a detailed characterization using different organic fractions in wastewater. *Journal of Membrane Science* **278**, 232–238. doi:10.1016/j.memsci.2005.11.006.

Sun, J., Hu, C., Zhao, K., Li, M., Qu, J. & Liu, H. 2018 Enhanced membrane fouling mitigation by modulating cake layer porosity and hydrophilicity in an electro-coagulation/oxidation membrane reactor (ECOMR). *Journal of Membrane Science* **550**, 72–79. doi:10.1016/j.memsci.2017.12.073.

Takeuchi, H. & Tanaka, H. 2020 Water reuse and recycling in Japan – history, current situation, and future perspectives. *Water Cycle* **1**, 1–12. doi:10.1016/j.watcyc.2020.05.001.

Tang, S., Wang, Z., Wu, Z. & Zhou, Q. 2010 Role of dissolved organic matters (DOM) in membrane fouling of membrane bioreactors for municipal wastewater treatment. *Journal of Hazardous Materials* **178**, 377–384. doi:10.1016/j.jhazmat.2010.01.090.

Thomas, R., Guillen-Burrieza, E. & Arafat, H. A. 2014 Pore structure control of PVDF membranes using a 2-stage coagulation bath phase inversion process for application in membrane distillation (MD). *Journal of Membrane Science* **452**, 470–480. doi:10.1016/j.memsci.2013.11.036.

Tian, J., Wu, C., Yu, H., Gao, S., Li, G., Cui, F. & Qu, F. 2018 Applying ultraviolet/persulfate (UV/PS) pre-oxidation for controlling ultrafiltration membrane fouling by natural organic matter (NOM) in surface water. *Water Research* **132**, 190–199. doi:10.1016/j.watres.2018.01.005.

Tlili, I. & Alkanhal, T. A. 2019 Nanotechnology for water purification: electrospun nanofibrous membrane in water and wastewater treatment. *Journal of Water Reuse and Desalination* **9** (3), 232–248. doi:10.2166/wrd.2019.057.

Wan, Y., Xie, P., Wang, Z., Wang, J., Ding, J., Dewil, R. & Van der Bruggen, B. 2021 Application of UV/chlorine pretreatment for controlling ultrafiltration (UF) membrane fouling caused by different natural organic fractions. *Chemosphere* **263**, 127993. doi:10.1016/j.chemosphere.2020.127993.

Wang, Y. & Tang, C. Y. 2011 Fouling of nanofiltration, reverse osmosis, and ultrafiltration membranes by protein mixtures: the role of inter-foulant-species interaction. *Environmental Science & Technology* **45** (15), 6373–6379. doi:10.1021/es2013177.

Wang, W., Yue, Q., Li, R., Bu, F., Shen, X. & Gao, B. 2018a Optimization of coagulation pre-treatment for alleviating ultrafiltration membrane fouling: the role of floc properties on Al species. *Chemosphere* **200**, 86–92. doi:10.1016/j.chemosphere.2018.02.114.

Wang, Z., Nan, J., Ji, X. & Yang, Y. 2018b Effect of the micro-flocculation stage on the flocculation/sedimentation process: the role of shear rate. *Science of the Total Environment* **633**, 1183–1191. doi:10.1016/j.scitotenv.2018.03.286.

Wang, W., Yue, Q., Guo, K., Bu, F., Shen, X. & Gao, B. 2019 Application of Al species in coagulation/ultrafiltration process: influence of cake layer on membrane fouling. *Journal of Membrane Science* **572**, 161–170. doi:10.1016/j.memsci.2018.11.014.

Wang, N., Li, X., Yang, Y., Zhou, Z., Shang, Y. & Zhuang, X. 2020 Photocatalysis-coagulation to control ultrafiltration membrane fouling caused by natural organic matter. *Journal of Cleaner Production* **265**, 121790. doi:10.1016/j.jclepro.2020.121790.

Wu, S., Hua, X., Miao, R., Ma, B., Hu, C., Liu, H. & Qu, J. 2020 Influence of floc charge and related distribution mechanisms of humic substances on ultrafiltration membrane behavior. *Journal of Membrane Science* **609**, 118260. doi:10.1016/j.memsci.2020.118260.

Wu, S., Hua, X., Ma, B., Fan, H., Miao, R., Ulbricht, M., Hu, C. & Qu, J. 2021 Three-dimensional analysis of the natural-organic-matter distribution in the cake layer to precisely reveal ultrafiltration fouling mechanisms. *Environmental Science & Technology* **55** (8), 5442–5452. doi:10.1021/acs.est.1c00435.

Xiong, X., Xu, H., Zhang, B., Wu, X., Sun, H., Wang, D. & Wang, Z. 2019 Floc structure and membrane fouling affected by sodium alginate interaction with Al species as model organic pollutants. *Journal of Environmental Sciences* **82**, 1–13. doi:10.1016/j.jes.2019.02.022.

Yao, M., Nan, J., Chen, T., Zhan, D., Li, Q., Wang, Z. & Li, H. 2015 Influence of flocs breakage process on membrane fouling in coagulation/ultrafiltration process – effect of additional coagulant of poly-aluminum chloride and polyacrylamide. *Journal of Membrane Science* **491**, 63–72. doi:10.1016/j.memsci.2015.05.018.

Yu, W., Zhang, D. & Graham, N. J. D. 2017 Membrane fouling by extracellular polymeric substances after ozone pre-treatment: variation of nano-particles size. *Water Research* **120**, 146–155. doi:10.1016/j.watres.2017.04.080.

Yu, W., Liu, T., Crawshaw, J., Liu, T. & Graham, N. 2018 Ultrafiltration and nanofiltration membrane fouling by natural organic matter: mechanisms and mitigation by pre-ozonation and pH. *Water Research* **139**, 353–362. doi:10.1016/j.watres.2018.04.025.

Yu, C., Gao, B., Wang, W., Xu, X. & Yue, Q. 2019 Alleviating membrane fouling of modified polysulfone membrane via coagulation pretreatment/ultrafiltration hybrid process. *Chemosphere* **235**, 58–69. doi:10.1016/j.chemosphere.2019.06.146.

Zhao, C., Song, T., Yu, Y., Qu, L., Cheng, J., Zhu, W., Wang, Q., Li, P. & Tang, W. 2020 Insight into the influence of humic acid and sodium alginate fractions on membrane fouling in coagulation-ultrafiltration combined system. *Environmental Research* **191**, 110228. doi:10.1016/j.envres.2020.110228.

First received 14 June 2021; accepted in revised form 18 August 2021. Available online 24 September 2021

Water Recycling and Low-Carbon Development doi: 10.2166/wrd.2021.042

Water quality deterioration of reclaimed water produced by reverse osmosis process in large pilot-scale distribution systems of different pipe materials

Xiyan Xu[a], Yong Cui[a,b,*], Zhangqing Wang[b], Hongyu Zhang[b], Caifeng Li[b] and Kanghua Yu[c]
[a] College of Chemistry and Chemical Engineering, Beijing Institute of Technology, Beijing 100081, China
[b] Beijing Yizhuang Environmental Technology Group Co. Ltd, Beijing 100176, China
[c] China Capital Management Co. Ltd., Beijing 100010, China
*Corresponding author. E-mail: cuiyong@bdawater.com

ABSTRACT

This work reports the effects of five pipe materials on reverse osmosis (RO)-reclaimed water quality in a large pilot-scale distribution system. These materials includes cast iron (CI), cast iron with cement-mortar lining (CML), stainless steel (SS), PVC, and PE. Long-term running tests for 96 hours are conducted with water quality parameters monitored online and analyzed offline. The results showed that red water appeared in CI pipe due to iron corrosion. The pH and TDS increased during the long-term test. Alkali–silica reaction in CML pipe led to a high increase of pH from 6.3 to 11.4, and TDS from 51 to 230 mg/L. Water quality deterioration was not observed in SS, PVC, and PE pipes. Residual chlorine decay occurred in all the five material pipes with the decay rate order of CI $\gg$ CML $>$ SS $\approx$ PVC $\approx$ PE. Ion concentration variation was also followed during the tests. Fe and Mn ions were detected in CI pipe and Ca, Si, Al, and S were detected in CML pipe. No detectable ion release was found in SS, PVC, and PE pipes. A kinetic model was postulated for the detected ion release with the mechanisms discussed in-depth.

Key words: circulation pipe system, pipe material, RO-reclaimed water, water quality

HIGHLIGHTS

- Reclaimed water quality deterioration is checked in five pipe materials.
- The tests are conducted in large scale reclaimed water pipe distribution systems.
- The results are obtained from 96 hours long-term running.
- A solid–liquid interface-based kinetic model is raised to describe the ion leaching.

1. INTRODUCTION

In the context of increasing water shortage, reclaimed water is gradually being seen as an important water resource in cities. Thus, the proportion of reclaimed water in total water resource consumption is rising yearly (Gao *et al.* 2007; Cherchi *et al.* 2019; T/CSES 2020). Reclaimed water is now mainly used for agricultural and garden irrigation, industry, environmental protection, entertainment, groundwater supplement, and power generation (Xu *et al.* 2016; Cherchi *et al.* 2019; Dery *et al.* 2019; Savchenko *et al.* 2019). According to different application purposes, various water treatment approaches are used for water reclamation, such as coagulation, ultrafiltration, ionic exchange, biodegradation, micro-flocculation, and reverse osmosis (RO) (Mouri *et al.* 2013; Hooper *et al.* 2014). RO method, always combining with affiliated steps like pressurization and degasification, can effectively remove most of the impurities and ions in water (Shenvi *et al.* 2015). Due to its high purification efficiency and cost-effectiveness, the RO process has been widely used for water reclamation in recent years. However, the highly purified water after RO reclamation is still corrosive since its stability falls out of the stable range. Meanwhile, the addition of disinfection agents can further increase the corrosivity. Thus, when RO-reclaimed water is transported through the distribution system, corrosion occurs on the inner pipe wall due to the reaction between the pipe materials and the reclaimed water. Consequently, a corrosion layer will form after long-term service which causes deterioration of the water quality (Nawrocki *et al.* 2010; Hu *et al.* 2018; Zhang *et al.* 2019). This brings risks for the users and elevates the maintenance cost of the water supply pipe network.

Previous works have studied the water quality variation caused by cast iron (CI) in various drinking water distribution systems. In the drinking water distribution network of Paris, France, the corrosion occurrence was studied in a full-scale spatio-temporal way to monitor water quality in CI pipes (Perrin *et al.* 2019). Sarin *et al.* (2004) reported on iron release from corroded CI pipes by dissolution of corrosion scales in drinking water distribution systems in the United States. To protect the CI pipe from corrosion, cement-mortar lined pipe (i.e., CML pipe) is used. However, cement-mortar lining can dissolve to release ions in corrosive reclaimed water and impact water quality (Song *et al.* 2016). Stainless steel (SS) is known for its high resistance toward corrosion, and it is thus widely used in highly corrosive environments (Du *et al.* 2016; Lindgren *et al.* 2018; Cui *et al.* 2019). Increasing efforts have been made to clarify the water quality and SS surface interaction, since corrosion still occurs on SS during long-term service (Cui *et al.* 2016a; Xu *et al.* 2019; Zhang *et al.* 2019; Xu *et al.* 2020). Plastic pipes, such as PVC and PE, are always used for short-distance water transportation due to their low sensitivity to electrochemical corrosion and cost-effectiveness (Lee *et al.* 2018), whereas their low mechanical strength restrains their application for long-distance water transportation.

Up to now, few studies have aimed at comparing the impact of different pipe materials on water deterioration in reclaimed water distribution systems (Wang *et al.* 2012; Lee *et al.* 2018; Xu *et al.* 2020). Given the increasing production and utilization of reclaimed water, further studies on the effect of different pipe materials on reclaimed water quality are necessitated. Previous studies either conducted lab-scale experiments with controlled conditions to check the water quality variation (Zhang *et al.* 2015; Song *et al.* 2016; Wang *et al.* 2018), or collected components serviced for a long time in real water distribution systems to analyze the corrosion scales and postulate the formation mechanisms (Jin *et al.* 2015; Cui *et al.* 2016b; Xu *et al.* 2019). The lab-scale experiments cannot exactly simulate the complex conditions in the real case. Meanwhile, the analysis of pipes from the real systems cannot provide continuous information on water deterioration. Thus, the kinetics of ion leaching during the deterioration cannot be obtained. Pilot-scale running tests cannot only easily control the conditions and obtain continuous data, but also simulate the real case to a large extent.

The aim of the current study is to compare the effect of different pipe materials on the deterioration of reclaimed water and discuss the relevant materials involved. For that purpose, RO-reclaimed water quality was monitored online and measured offline in a large pilot-scale circulation pipe system made of pipe materials that included CI, CML, SS, PVC, and PE. The pH, TDS, residual chlorine, and concentration of ions were followed during long-term running tests for 96 hours, and the kinetics of iron leaching and mechanisms for the variation of the water parameters were discussed in-depth. The results from the current work are useful for the pipe material selection in reclaimed water distribution systems.

2. MATERIALS AND METHODS

2.1. Experimental equipment

The pilot-scale circulation pipe system was composed of pipelines, circulation pump, and buffer tank fixed by a steel supporter as shown in Figure 1. The parallel pipelines contained five materials (CI, CML, SS, PVC, and PE). The pipeline arrangement of the materials was the same. The diameter of all the pipes was DN100 and the length was 120 m. A 30 L water tank was placed as the buffer of the system to avoid excessive fluctuation of hydraulic pressure.

2.2. Reclaimed water

The high-quality reclaimed water tested in this study was supplied from a reclaimed water plant in Beijing, China. This plant used a double membrane process (micro filtering + reverse osmosis, i.e., MF + RO) to treat the outlet water from a nearby wastewater treatment plant. A BW30-365FR type RO membrane was used, which can remove above 99.5% of minerals from the water. The reclaimed water quality parameters are displayed in Table 1 and the corresponding drinking water quality (acquired from the tap in the reclaimed plant) is also shown for comparison. Compared with drinking water, COD and BOD in reclaimed water were higher, but other parameters, including pH, TDS, conductivity, calcium, general hardness, and general alkalinity, were much lower. That is because the MF and RO process can remove most of the ions and suspended solids in water (Shenvi *et al.* 2015; Liu *et al.* 2018).

2.3. The running tests and sampling

The pilot-scale pipe systems were filled with reclaimed water by pumping and then sealed. The water circulation was initiated with the help of a circulating pump and kept at a stable flow rate of 1.4 m/s. During the 96 hours of tests, pH, TDS, residual chlorine, and conductivity were monitored online. Meanwhile, water sampling was conducted and saved in a fridge for

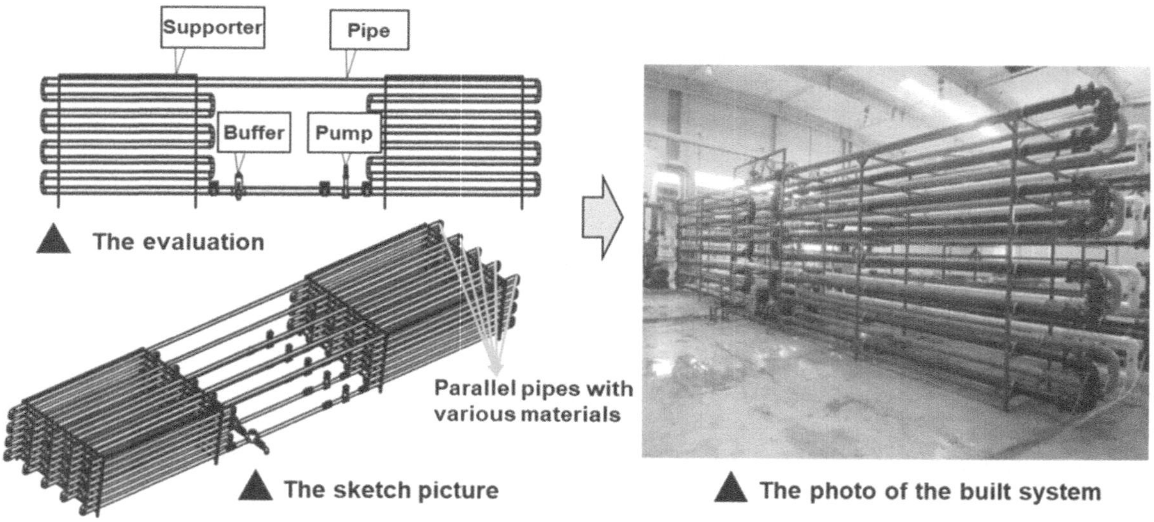

Figure 1 | The pilot-scale circulation pipe distribution system of reclaimed water.

Table 1 | Comparison of water quality for reclaimed water and drinking water

Water quality parameter	Reclaimed water	Drinking water
pH	6.0	7.2
COD (mg/L)	2.09	0.8
BOD_5 (mg/L)	0.57	0.24
TDS (mg/L)	14.6	252
Conductivity (μS/cm)	26.2	400
Calcium (Ca) (mg/L)	0.15	50.7
General hardness ($CaCO_3$) (mg/L)	1.06	227
General alkalinity ($CaCO_3$) (mg/L)	7.9	164

further offline analysis. The offline analysis was carried out within 24 hours. A small amount of nitric acid was added into the water samples to resolve the soluble solids before analysis.

The pH value, TDS, residual chlorine, and conductivity were detected by water quality analyzers, including PHS-3C, DZS-708, DGB-402, and DDSJ-319 L meters (INESA Scientific Instrument Co., Ltd, China). The ions' concentration, including Ca, Si, Al, Fe, Mn, and S were measured by inductively coupled plasma emission spectrometer (PerkinElmer, USA).

3. RESULTS AND DISCUSSION

In the pilot-scale circulating pipe system, reclaimed water circled along the CI, CML, SS, PVC, and PE pipes simultaneously. Before the tests, the pipes were washed by high-speed flow to remove attached impurities. The pH value has been proven to adjust the chemical balance in pipe water and thus impacts greatly on the dissolution and precipitation of ions in water; TDS is a comprehensive parameter that reflects the solids from the pipe inner wall dissolved into water. It can be seen as an indicator for water quality deterioration; residual chlorine is always added into the reclaimed water to restrain the microbial growth, so its decay rate significantly impacts on the water quality during distribution. In that respect, to clarify the impact of the five pipe materials on reclaimed water quality during the long-term tests, the pH value, TDS, and residual chlorine were followed and studied.

3.1. Variations of pH in pipes with different materials

The variation of pH values in different pipe materials are shown in Figure 2(a). In CI pipe, pH increased from 6.2 to 6.8 quickly in the initial 8 hours and then remained steady. In the initial stage, the pH increase is due to the corrosion of the

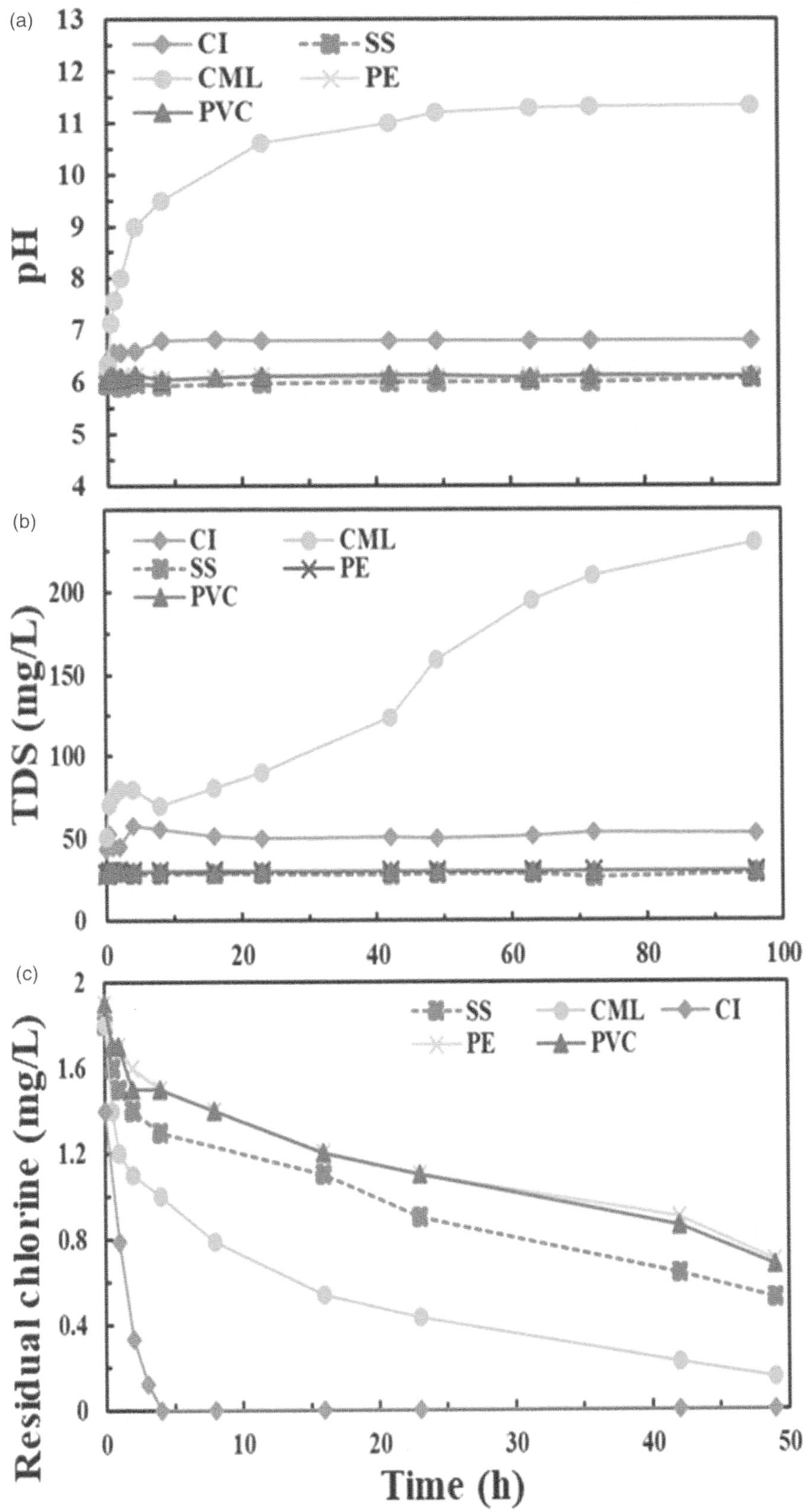

Figure 2 | The variation of (a) pH, (b) TDS, and (c) residual chlorine concentration in pipes with different materials during the long-term running tests.

inner wall of CI pipe in contact with reclaimed water (Cui *et al.* 2016b; Zhang *et al.* 2019). The dissolved oxygen in water reacted with the iron material as seen in Equations (1)–(4). With the consumption of dissolved oxygen, the water becomes anoxic and Fe_3O_4 could form to restrain further corrosion as shown in Equation (5):

$$Fe \rightarrow Fe^{2+} + 2e^- \tag{1}$$

$$O_2 + 2H_2O + 4e^- \rightarrow 4OH^- \tag{2}$$

$$Fe^{2+} + 2OH^- \rightarrow Fe(OH)_2 \tag{3}$$

$$4Fe(OH)_2 + O_2 + 2H_2O \rightarrow 4Fe(OH)_3 \tag{4}$$

$$Fe(OH)_2 + 2Fe(OH)_3 \rightarrow Fe_3O_4 + 4H_2O \tag{5}$$

In CML pipe, the pH value of water increased greatly from 6.34 to 9 in the first 4 hours. From 4 to 42 hours, pH continued to increase at a slower rate to 11 at 42 hours. Then, it became relatively stable. During the long-term test, alkali–silica in the cement mortar reacted with water. The cement mortar contained reactants including silica in the solid state, potassium and/or sodium hydroxide and possibly calcium (Saouma *et al.* 2015). When subsequent cement hydrate such as ettringite was formed, the corresponding counter ion of alkalis became hydroxide (OH^-), leading to the increase of water pH in the CML pipe. Regarding SS, PVC, and PE pipes, the pH was almost stable during 96 hour tests, showing the corrosion resistance of these pipes (Asri *et al.* 2017).

3.2. Variation of TDS in different material pipes

Figure 2(b) shows the TDS variation in different pipes. In CI pipe, TDS increased from 44 to 58 mg/L in the first 4 hours, which is on account of mineral ions dissolving from the CI, as shown in Equations (3) and (4). After 4 hours, a slight decline of TDS was observed. The protective coating effect of the Fe_3O_4 formed on the inner wall of the CI pipe, as indicated by Equation (5), played a role in restraining the increase of TDS. Meanwhile, precipitation of some ions may occur as the result of the pH value increase, causing the decline of TDS. Regarding the CML pipe, in the first stage of 4 hours, the hydrate, hydroxide, and ions leached from the inner pipe wall into water because of the alkali–silica reaction. Thus, TDS increased from 51 to 80 mg/L. Then, a decline of TDS down to around 70 mg/L occurred until 8 hours due to the increase of pH value. From then on, TDS kept growing to 230 mg/L at the end of the 96 hour tests at high rate. This is because, on one hand, OH^- attacked aggregate silanol and siloxane groups and formed dissolved silica ions; on the other hand, cations in pore solution of cement mortar combined with dissolved silica ions to result in new hydrates, and increasing ions thus dissolved from cement mortar into the reclaimed water (Kleib *et al.* 2018; Yazici *et al.* 2019). In SS, PVC, and PE pipes, TDS was shown to be quite stable during the 96 hour tests.

3.3. Variation of residual chlorine in different material pipes

Figure 2(c) shows that residual chlorine decayed in all of the five kinds of pipes with different rates due to different reactions in these pipes (Hallam *et al.* 2002). The decay rates of the residual chlorine concentration were as the following order: $CI \gg CMI > SS \approx PVC \approx PE$.

In CI pipe, 1.8 mg/L residual chlorine was exhausted completely at the fastest rate, within 4 hours. During this period, the residual chlorine in the reclaimed water was consumed by the oxidation and reduction reactions with CI as indicated in Equations (1) and (6) (Zhang *et al.* 2019):

$$6Fe^{2+} + 3ClO^- + 3H_2O \rightarrow 2Fe(OH)_3 + 4Fe^{3+} + 2Cl^- \tag{6}$$

The residual chlorine decay was slower in CML pipe. During the long-term test, the concentration of residual chlorine decreased continuously from 1.8 mg/L to 0.15 mg/L. The rough inner wall of CML pipe provided anchoring sites for microbials so that the residual chlorine was consumed for disinfection (Li *et al.* 2019). However, the Fe-related reactions in CI pipe did not occur in CML pipe, resulting in a slower residual chlorine decay. In SS, PVC, and PE pipes, the variation of the residual chlorine concentration was even slower due to their inertia in chlorine-involving reactions. Residual chlorine decreased from 1.8 mg/L to 0.52 mg/L, 0.68 mg/L, and 0.7 mg/L in SS, PVC, and PE pipes, respectively.

Residual chlorine decay curve fitted well with the second order kinetic equation. The kinetic rate constants with the correlation coefficients ($R^2 > 0.90$) are shown in Table 2. As can be seen, the rate constant of residual decay in CI pipe was over 69 times higher than other pipes. The rate constant of CML pipe ranked in second place followed by SS pipe. The constants in PVC and PE were similar and were all very low since the pipe materials were both plastic-based.

3.4. Ion concentration increase in different material pipes

The variation of pH value, TDS, and residual chlorine analysis indicated that the water quality was affected to different extents in various pipe materials during the long-term running tests. This variation was on account of ion leaching from the pipe inner wall. The concentration of ions in the reclaimed water at 49 hours are shown in Table 3 during the long-term running tests in the five pipe materials. It can be seen that Fe and Mn ions were detected in CI pipe, and Ca and Si were found in CML pipe. No Fe ions were detected in SS, PVC, and PE pipes. The results so far showed that SS, PVC, and PE pipes were relatively stable and did not cause obvious deterioration of reclaimed water quality during the long-term tests, whereas CI and CML pipes showed a more remarkable influence on the water quality. Thus, the ion concentration increases in CI and CML pipes were studied in-depth.

3.4.1. Ion concentration increase in water sample of CI and CML pipe

The variation of total Fe and Mn concentration in CI pipe is shown in Figure 3(a). The Fe ion concentration reached around 50 mg/L within the first hour and fluctuated greatly afterwards during the long-term test. The manganese ion was lower, and it kept growing continuously to a balance concentration of 0.25 mg/L at around 48 hours. Red water was observed in CI pipe during the long-term test. The corrosivity of reclaimed water led to ion leaching from the CI pipe inner wall. The leached ions formed precipitation gradually at the function of dissolved oxygen and increasing pH value (Equations (1)–(4)). The precipitation fell off from the pipe inner wall by the water flow and formed the turbid red water during the long-term running test (Cui *et al.* 2016b).

The concentrations of Ca, Si, Al, and S ions during the long-term tests are shown in Figure 3(b). The ion concentrations increased in the beginning and reached balance concentration of 41.3 mg/L, 12.2 mg/L, 1.21 mg/L, and 1.25 mg/L for Ca, Si, Al, and S ions at around 64 hours. The appearance of these ions indicated that components of cement dissolved into water. The components included hydrated calcium silicate, calcium hydroxide, calcium aluminate hydrate, and calcium sulfoaluminate hydrate. The following reactions occurred resulting in the increase of ion concentration (Kleib *et al.* 2018). The

Table 2 | Second order kinetic rate constants of residual decay in reclaimed water in different pipe materials

	CI	CML	SS	PVC	PE
k (mg/(L min))	2.443	0.0352	0.0163	0.0085	0.0071
R^2	0.91	0.92	0.99	0.97	0.90

Table 3 | Ion concentration in different pipe materials at 49 hours

Ions (mg/L)	CI	SS	CML	PVC	PE
Fe	50	–[a]	–	–	–
Mn	0.24	–	–	–	–
Cr	–	–	–	–	–
Ni	–	–	–	–	–
Ca	–	–	26.16	–	–
Si	–	–	9.32	–	–
Al	–	–	1.93	–	–
Sulfate	–	–	1.18	–	–

[a]–, not detected.

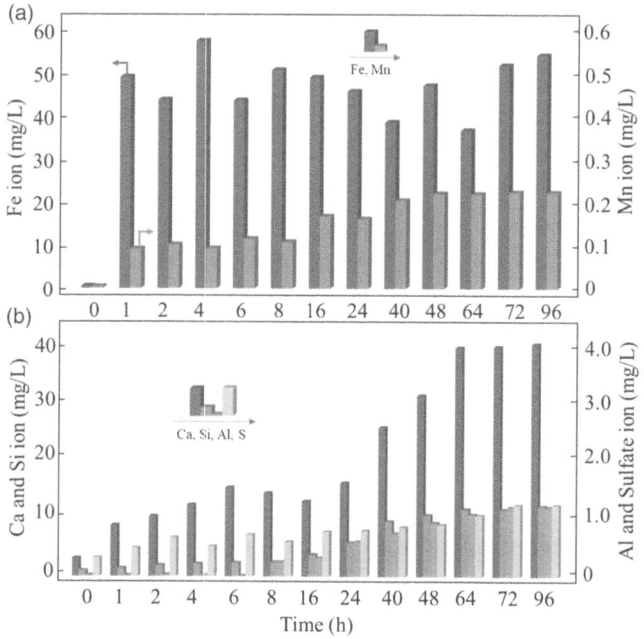

Figure 3 | Iron concentration in CI (a) and CML (b) pipes during the 96 hours' long-term running tests.

dissolution of these components raised pH and TDS.

$$Ca(OH)_2 \rightarrow Ca^{2+} + 2OH^- \tag{7}$$

$$CaO \cdot SiO_2 \cdot YH_2O + H_2O \rightarrow Ca^{2+} + 2OH^- + SiO_3^- + H_2O \tag{8}$$

$$3CaO \cdot Al_2O_3 \cdot 6H_2O + H_2O \rightarrow Ca^{2+} + 2OH^- + AlO_2^- + H_2O \tag{9}$$

$$3CaO \cdot Al_2O_3 \cdot 3CaSO_4 \cdot 32H_2O + H_2O \rightarrow Ca^{2+} + 2OH^- + AlO_2^- + SO_4^{2-} + H_2O \tag{10}$$

3.4.2. Kinetic model for ion leaching in CI and CML pipes

The processes of different ions leaching from the inner wall of CI and CML pipe into the reclaimed water were complicated. In the current study, a simplified apparent kinetic model including three stages was used for describing ion leaching (Figure 4): (1) the elements in the crystal on the pipe inner wall dissolved into the solid–liquid interface layer and became ions; (2) ions were transported in the solid–liquid interface layer toward the water body due to the concentration difference; and (3) ions continuously moved into the reclaimed water body and the ion concentration in the water body increased.

The rate-determining step in the process was the ions' transportation in the interface layer (stage 2). Based on Fick's law, the following equations can be given to describe the ion leaching:

$$J = D \cdot \frac{dC}{dx} \tag{11}$$

$$J = \frac{dM}{Sdt} \tag{12}$$

$$\frac{dC}{dx} = \frac{C_s - C}{h} \tag{13}$$

$$M = C \cdot V \tag{14}$$

where J is the diffusion flux; D is the diffusion coefficient; C_s is the dissolved equilibrium concentration of ion; C is the ion concentration in the reclaimed water body at time t; S is the surface area of the pipe inner wall; M is the ion mass passing

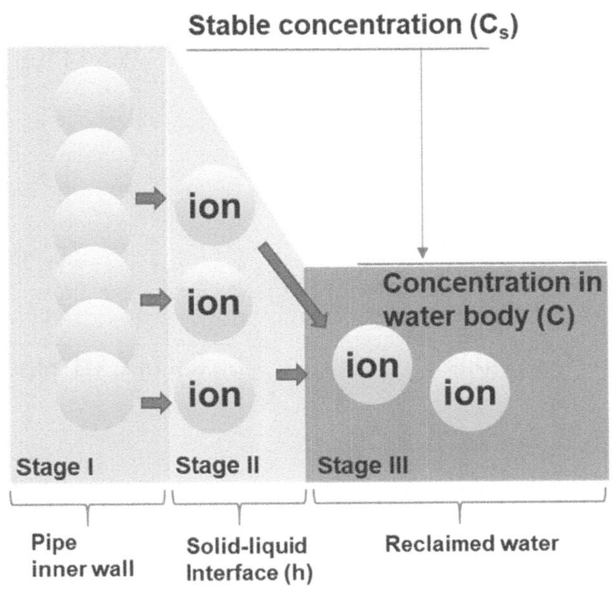

Figure 4 | Ion leaching process from pipe inner wall into reclaimed water.

through the unit area S within a time t; h is the thickness of the solid–liquid interface layer; and V is the volume of water body. This model was based on the following assumptions: (a) pipe inner wall started quickly to dissolve; (b) no ion concentration difference existed in the reclaimed water body; (c) the ion concentration in the layer near the surface of the solid pipe inner wall reached saturation; and (d) the critical water layer is a stable water layer with a stable thickness of h.

Equations (11)–(14) can be comprehensively rewritten as Equation (15):

$$\frac{\mathrm{d}C}{\mathrm{d}t} = \frac{DS(C_s - C)}{Vh} \tag{15}$$

Since D, S, V, and h are all constants, DS/Vh can be substituted by k', thus, Equation (15) can be rewritten as Equation (16):

$$\frac{\mathrm{d}C}{\mathrm{d}t} = \mathrm{k}'(C_s - C) \tag{16}$$

The kinetic plots for Fe and Mn ion leaching in CI pipe and Ca, Si, Al, and S leaching in CML pipe are shown in Figure 5. The corresponding k' values and R^2 are also included. Looking at the R^2, Mn ion in CI pipe and Si and Al ions in CML pipe fitted well with the kinetic model ($R^2 > 0.9$). However, Fe ion in CI pipe did not fit the kinetic model used in the current study given the low R^2 value. This was due to the complicated behavior of Fe corrosion and Fe ion leaching during the long-term test, and thus resulting in the great fluctuation of Fe ion concentration (Figure 3). This required further study to postulate a more detailed and precise model for Fe ion leaching during long-term pilot-scale tests.

4. CONCLUSIONS

The current work studied the water quality deterioration in five different pipe materials for the transportation of RO-reclaimed water in large pilot-scale distribution systems. Alkali–silica reaction took place in the CML pipe, leading to a great increase of pH and TDS values. The pH and TDS value fluctuation extent in CI pipe ranked second among the five materials. These two values were quite stable in SS, PVC, and PE pipes, showing their corrosion resistance in RO-reclaimed water. Residual chlorine decay was observed in all the five pipe materials according to an order of C I $\gg$ CMI $>$ SS $\approx$ PVC $\approx$ PE. The residual chlorine decay fitted with the second order kinetic equation with the decay rate being 2.443, 0.0352, 0.0163, 0.0085, and 0.0071 mg/(L min), respectively. No detectable ion leaching was observed in SS, PVC, and PE pipes. Fe and Mn ions were found in CI pipe and Ca, Si, Al, and S ions were detected in CML pipe. A simplified kinetic model was postulated to

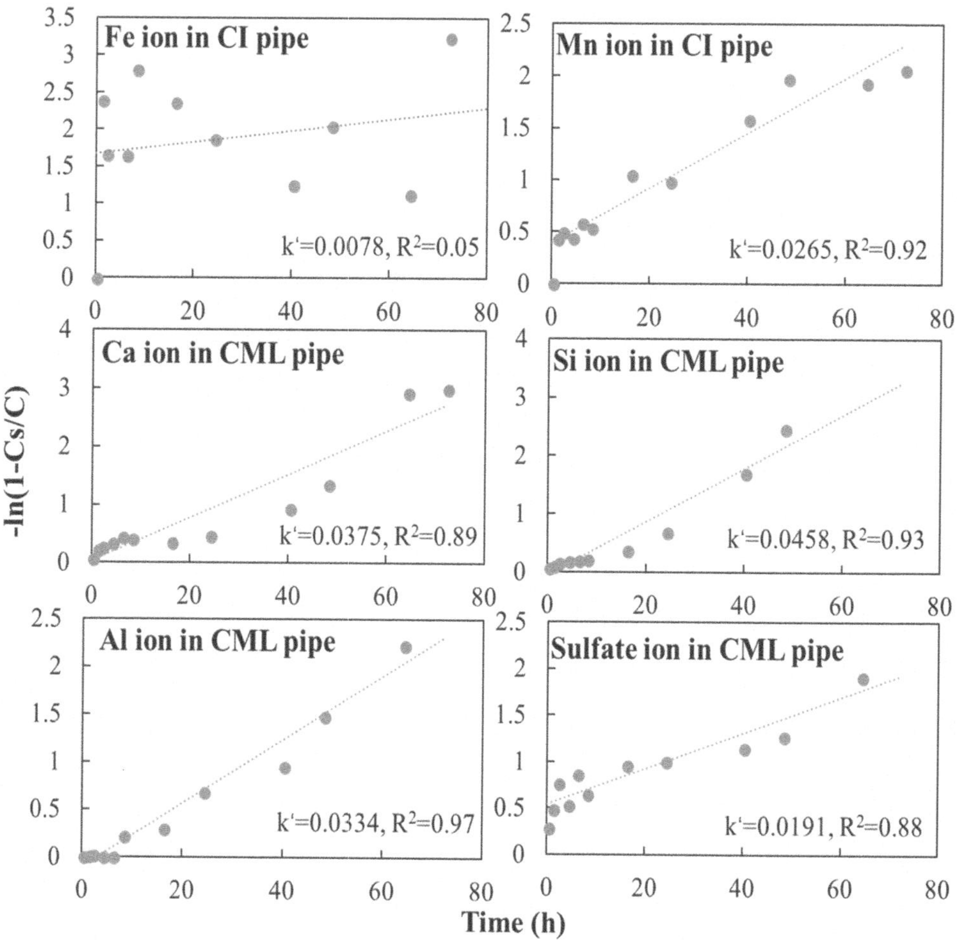

Figure 5 | Kinetic plots of ion leaching.

describe the ion leaching process. This model did not fit with the Fe leaching since the Fe ion fluctuated greatly during the long-term tests.

The results from this work indicated that SS, PVC, and PE pipes were resistant to corrosion in the RO-reclaimed water and caused a relatively low extent of water quality deterioration. Future studies are encouraged to consider comprehensively the ion release, residual chlorine decay as well as the mechanical strength. That may provide further knowledge on the pipe material selection for reclaimed water transportation, helping to achieve a cleaner production of reclaimed water.

ACKNOWLEDGEMENTS

We are grateful for the support of China Postdoctoral Science Foundation (No. 2018M631495), National Natural Science Foundation of China (51808312, 51879139).

DATA AVAILABILITY STATEMENT

Data cannot be made publicly available; readers should contact the corresponding author for details.

REFERENCES

Asri, N. F., Husaini, T., Sulong, A. B., Majlan, E. H. & Daud, W. R. W. 2017 Coating of stainless steel and titanium bipolar plates for anticorrosion in PEMFC: a review. *International Journal of Hydrogen Energy* **42** (14), 9135–9148.

Cherchi, C., Kesaano, M., Badruzzaman, M., Schwab, K. & Jacangelo, J. G. 2019 Municipal reclaimed water for multi-purpose applications in the power sector: a review. *Journal of Environmental Management* **236**, 561–570.

Cui, Y., Liu, S., Smith, K., Hu, H., Tang, F., Li, Y. & Yu, K. 2016a Stainless steel corrosion scale formed in reclaimed water: characteristics, model for scale growth and metal element release. *Journal of Environmental Sciences* **48**, 79–91.

Cui, Y., Liu, S., Smith, K., Yu, K., Hu, H., Jiang, W. & Li, Y. 2016b Characterization of corrosion scale formed on stainless steel delivery pipe for reclaimed water treatment. *Water Research* **88**, 816–825.

Cui, Z., Chen, S., Dou, Y., Han, S., Wang, L., Man, C., Wang, X., Chen, S., Cheng, Y. F. & Li, X. 2019 Passivation behavior and surface chemistry of 2507 super duplex stainless steel in artificial seawater: influence of dissolved oxygen and pH. *Corrosion Science* **150**, 218–234.

Dery, J. L., Rock, C. M., Goldstein, R. R., Onumajuru, C., Brassill, N., Zozaya, S. & Suri, M. R. 2019 Understanding grower perceptions and attitudes on the use of nontraditional water sources, including reclaimed or recycled water, in the semi-arid Southwest United States. *Environmental Research* **170**, 500–509.

Du, D., Chen, K., Lu, H., Zhang, L., Shi, X., Xu, X. & Andresen, P. L. 2016 Effects of chloride and oxygen on stress corrosion cracking of cold worked 316/316l austenitic stainless steel in high temperature water. *Corrosion Science* **110**, 134–142.

Gao, X., Wu, X. Q., Zhang, Z. E., Guan, H. & Han, E. H. 2007 Characterization of oxide films grown on 316l stainless steel exposed to H_2O_2-containing supercritical water. *Journal of Supercritical Fluids* **42** (1), 157–163.

Hallam, N. B., West, J. R., Forster, C. F., Powell, J. C. & Spencer, I. 2002 The decay of chlorine associated with the pipe wall in water distribution systems. *Water Research* **36** (14), 3479–3488.

Hooper, T., Cooper, P., Hunt, A. & Austen, M. 2014 A methodology for the assessment of local-scale changes in marine environmental benefits and its application. *Ecosystem Services* **8**, 65–74.

Hu, J., Dong, H., Xu, Q., Ling, W., Qu, J. & Qiang, Z. 2018 Impacts of water quality on the corrosion of cast iron pipes for water distribution and proposed source water switch strategy. *Water Research* **129**, 428–435.

Jin, J., Wu, G. & Guan, Y. 2015 Effect of bacterial communities on the formation of cast iron corrosion tubercles in reclaimed water. *Water Research* **71**, 207–218.

Kleib, J., Aouad, G., Louis, G., Zakhour, M., Boulos, M., Rousselet, A. & Bulteel, D. 2018 The use of calcium sulfoaluminate cement to mitigate the alkali silica reaction in mortars. *Construction and Building Materials* **184**, 295–303.

Lee, H., Rasheed, U. & Kong, M. 2018 A study on the comparison of corrosion in water supply pipes due to tap water (TW) and reclaimed water (RW). *Water* **10** (4), 496.

Li, R. A., McDonald, J. A., Sathasivan, A. & Khan, S. J. 2019 Disinfectant residual stability leading to disinfectant decay and by-product formation in drinking water distribution systems: a systematic review. *Water Research* **153**, 335–348.

Lindgren, M., Huttunen-Saarivirta, E., Peltola, H., Romu, J., Sarikka, T., Hanninen, H. & Pohjanne, P. 2018 Crevice corrosion of stainless steels 904l, 2205, and 2507 in high-temperature sulfuric acid solution containing chlorides: influence of metal cations. *Corrosion* **74** (2), 225–240.

Liu, R., Raman, A. K. Y., Shaik, I., Aichele, C. & Kim, S.-J. 2018 Inorganic microfiltration membranes incorporated with hydrophilic silica nanoparticles for oil-in-water emulsion separation. *Journal of Water Process Engineering* **26**, 124–130.

Mouri, G., Takizawa, S., Fukushi, K. & Oki, T. 2013 Estimation of the effects of chemically-enhanced treatment of urban sewage system based on life-cycle management. *Sustainable Cities and Society* **9**, 23–31.

Nawrocki, J., Raczyk-Stanislawiak, U., Swietlik, J., Olejnik, A. & Sroka, M. J. 2010 Corrosion in a distribution system: steady water and its composition. *Water Research* **44** (6), 1863–1872.

Perrin, Y., Bouchon, D., Delafont, V., Moulin, L. & Hechard, Y. 2019 Microbiome of drinking water: a full-scale spatio-temporal study to monitor water quality in the Paris distribution system. *Water Research* **149**, 375–385.

Saouma, V. E., Martin, R. A., Hariri-Ardebili, M. A. & Katayama, T. 2015 A mathematical model for the kinetics of the alkali-silica chemical reaction. *Cement and Concrete Research* **68**, 184–195.

Sarin, P., Snoeyink, V. L., Bebee, J., Jim, K. K., Beckett, M. A., Kriven, W. M. & Clement, J. A. 2004 Iron release from corroded iron pipes in drinking water distribution systems: effect of dissolved oxygen. *Water Research* **38** (5), 1259–1269.

Savchenko, O. M., Kecinski, M., Li, T. & Messer, K. D. 2019 Reclaimed water and food production: cautionary tales from consumer research. *Environmental Research* **170**, 320–331.

Shenvi, S. S., Isloor, A. M. & Ismail, A. F. 2015 A review on RO membrane technology: developments and challenges. *Desalination* **368**, 10–26.

Song, Y., Tian, Y., Zhao, X., Guo, H. & Zhang, H. 2016 Corrosion process of ductile iron with cement mortar linings as coatings in reclaimed water. *International Journal of Electrochemical Science* **11** (8), 7031–7047.

T/CSES 2020 *Guidelines for Water Reuse – Reclaimed Water Classification and Marking*. Chinese Society for Environmental Sciences, China.

Wang, H., Hu, C., Hu, X., Yang, M. & Qu, J. 2012 Effects of disinfectant and biofilm on the corrosion of cast iron pipes in a reclaimed water distribution system. *Water Research* **46** (4), 1070–1078.

Wang, H., Hu, C., Zhang, S., Liu, L. & Xing, X. 2018 Effects of O-3/Cl-2 disinfection on corrosion and opportunistic pathogens growth in drinking water distribution systems. *Journal of Environmental Sciences* **73**, 38–46.

Xu, M., Bai, X., Pei, L. & Pan, H. 2016 A research on application of water treatment technology for reclaimed water irrigation. *International Journal of Hydrogen Energy* **41** (35), 15930–15937.

Xu, X., Liu, S., Liu, Y., Smith, K. & Cui, Y. 2019 Corrosion of stainless steel valves in a reverse osmosis system: analysis of corrosion products and metal loss. *Engineering Failure Analysis* **105**, 40–51.

Xu, X., Liu, S., Liu, Y., Smith, K., Wang, X., Li, J., Ma, Z., Wang, Z. & Cui, Y. 2020 Water quality induced corrosion of stainless steel valves during long-term service in a reverse osmosis system. *Journal of Environmental Sciences* **89**, 218–226.

Yazici, H., Beglarigale, A., Felekoglu, K. T. & Turkel, S. 2019 Comparing the alkali-silica reaction mitigation potential of admixtures by using different accelerated test methods. *Construction and Building Materials* **197**, 597–614.

Zhang, H., Tian, Y., Wan, J. & Zhao, P. 2015 Study of biofilm influenced corrosion on cast iron pipes in reclaimed water. *Applied Surface Science* **357**, 236–247.

Zhang, H., Tian, Y., Kang, M., Chen, C., Song, Y. & Li, H. 2019 Effects of chlorination/chlorine dioxide disinfection on biofilm bacterial community and corrosion process in a reclaimed water distribution system. *Chemosphere* **215**, 62–73.

First received 31 May 2021; accepted in revised form 17 August 2021. Available online 25 October 2021

Water Recycling and Low-Carbon Development doi: 10.2166/wrd.2021.040

Characterization of bacterial fluorescence: insight into rapid detection of bacteria in water

Yu Mao [a], Xiao-Wen Chen [a], Zhuo Chen [a,*], Gen-Qiang Chen [a], Yun Lu [a], Yin-Hu Wu [a] and Hong-Ying Hu [a,b]

[a] Environmental Simulation and Pollution Control State Key Joint Laboratory, State Environmental Protection Key Laboratory of Microorganism Application and Risk Control (SMARC), Beijing Laboratory for Environmental Frontier Technologies, School of Environment, Tsinghua University, Beijing 100084, China
[b] Research Institute for Environmental Innovation (Suzhou), Tsinghua, Suzhou 215163, China
*Corresponding author. E-mail: zhuochen@mail.tsinghua.edu.cn

YM, 0000-0003-0241-9545; ZC, 0000-0001-7799-5163

ABSTRACT

Microbial contamination is one of the main risks affecting water safety. Traditional microbial detection methods tend to be time-consuming and labor-intensive. Thus, this study investigated a potential rapid and simple method for bacterial detection in water by excitation–emission matrix (EEM) fluorescence spectroscopy. Particularly, bacterial intrinsic fluorophores were divided into three regions, namely Region A (amino acids), Region N (NAD(P)H) and Region F (flavins). Afterwards, fluorescence characteristics of four pure bacterial species (*Bacillus subtilis*, *Staphylococcus aureus*, *Escherichia coli* and *Pseudomonas aeruginosa*) as well as indigenous bacteria in secondary effluent from two water reclamation plants were evaluated via fluorescence regional integration (FRI). Correlation analysis between fluorescence intensity (FI) integral and bacterial concentration was conducted, and principal component analysis (PCA) was applied to distinguish the fluorescence spectra of different bacteria. The results showed that most of the bacterial autofluorescence was emitted by amino acids and the FI integral of flavins had a good linear relationship ($R^2 > 0.9$) with bacterial concentration. PCA could distinguish varied bacterial species and bacteria from different secondary effluents. This study indicated that FRI was helpful for the characterization of bacterial fluorescence and the quantification of bacteria in water.

Key words: bacterial detection, fluorescence regional integration, fluorescence spectroscopy, principal component analysis

HIGHLIGHTS

- Fluorescence regional integration was used to characterize bacterial intrinsic fluorescence.
- Most of the bacterial intrinsic fluorescence was emitted by amino acids.
- Flavin fluorescence was well correlated with bacterial concentration.

GRAPHICAL ABSTRACT

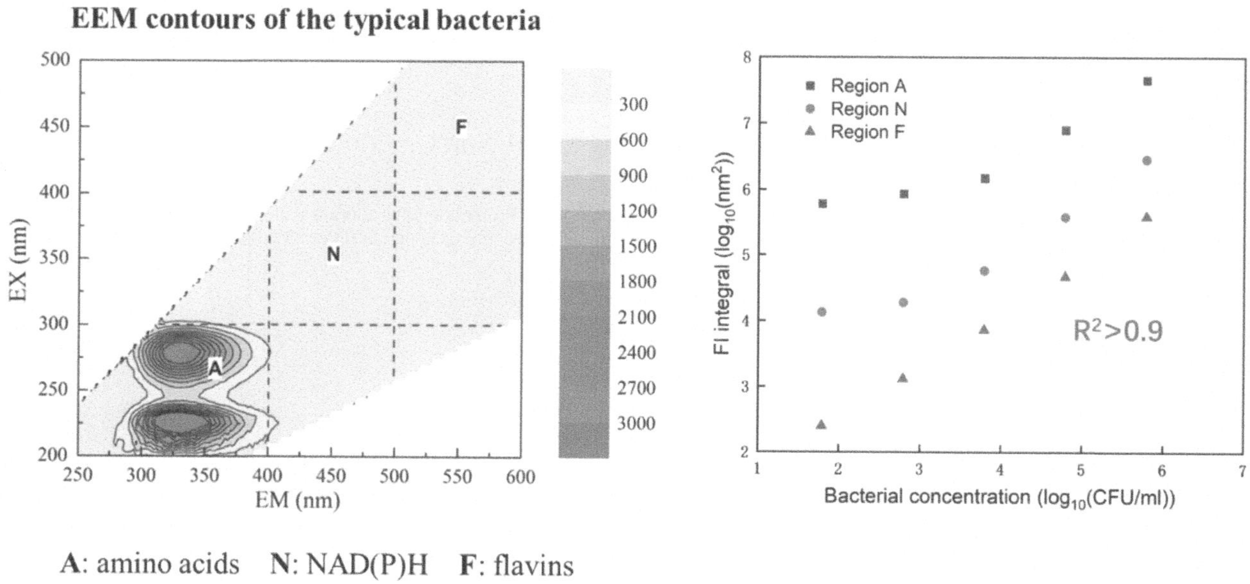

1. INTRODUCTION

Clean water is vital to public health as well as the sustainable development of society and economy (UN 2015; Shi *et al.* 2020). However, with the rapid population growth and the acceleration of urbanization and industrialization, water pollution is increasingly serious and has caused many detrimental health problems around the world (Clemens *et al.* 2017; Budnik & Casteleyn 2019; Chen *et al.* 2020). According to the fourth edition of the World Health Organization's (WHO) Guidelines for drinking-water quality, most of the evident water-related health problems are caused by microbial contamination (WHO 2011). Microorganisms in water include bacteria, viruses and protozoa, which may not only lead to the occurrence of water-borne diseases, but also pose other risks such as pipeline corrosion and membrane fouling (Matin *et al.* 2011; Enning & Garrelfs 2014; Luo *et al.* 2021). To efficiently prevent these adverse effects, it is essential to detect microbial contamination in water.

Culture-based methods (e.g., plate counting technique) have been widely used for microbial detection since the last century. However, the detection process is quite time-consuming and labor-intensive. Moreover, the existence of viable but nonculturable microorganisms may lead to the underestimation of the actual number of microorganisms (Rollins & Colwell 1986). Besides the traditional culture-based methods, emerging methods such as molecular methods (e.g., polymerase chain reaction), flow cytometry (FCM) and adenosine triphosphate (ATP) bioluminescence are increasingly used for microbial detection (Rajapaksha *et al.* 2019). Polymerase chain reaction (PCR) is a DNA amplification technique which can provide reliable results in several hours owing to its high sensitivity, but viable and nonviable microorganisms could not be distinguished since both types of DNA can be amplified (Cangelosi & Meschke 2014). In recent years, propidium monoazide (PMA) has been used to prevent the amplification of DNA from nonviable microorganisms (Banihashemi *et al.* 2015). Although PMA enables the detection of viable microorganisms, the detection process is quite complicated and includes PMA staining and DNA extraction. Besides, FCM is an optical detection method which analyzes the scatter light and/or fluorescence emitted by suspended microbial particles (Safford & Bischel 2019). The absolute number of microorganisms can be determined in very short times by this method. However, sample pretreatment and fluorescent staining may be needed, and the cost of this method is relatively expensive. In addition, ATP bioluminescence is a biochemical method which detects microbial activity since ATP exists in all living cells (Bottari *et al.* 2015). Although the detection process can be finished in several minutes, biotic and abiotic ATP could not be distinguished and some matrix effects may interfere with the determination or inhibit luminescence (Ntziachristos *et al.* 2005). Therefore, a rapid, simple, reliable and cost-effective method for microbial detection is urgently demanded.

Fluorescence spectroscopy is a fast and noninvasive technique which has been widely used for the characterization of organic matters and has a broad prospect in the real-time online detection of microorganisms (Carstea *et al.* 2016). Live bacteria contain many intrinsic fluorophores such as aromatic amino acids and coenzymes, which exhibit fluorescence at specific excitation and emission wavelengths (Ammor 2007). The relationship among excitation wavelengths, emission wavelengths and fluorescence intensity (FI) can serve as an intrinsic bacterial fingerprint. Early studies on bacterial fluorescence remained at a qualitative level, which focused on the identification and classification of bacterial species (Giana *et al.* 2003; Sohn *et al.* 2009). In recent years, fluorescence spectroscopy is increasingly used for the quantitative detection of bacteria in water. However, most relevant studies were based on tryptophan-like fluorescence, which was determined at a single excitation and emission wavelength (Baker *et al.* 2015; Bridgeman *et al.* 2015; Sorensen *et al.* 2015). A study of bacterial detection in groundwater compared the detection capability among a single excitation and emission wavelength pair, two-dimensional emission spectra and three-dimensional excitation–emission matrix (EEM), concluding that better quantification could be achieved based on partial least squares analysis of EEM and the detection limit was as low as 10 CFU/ml (Nakar *et al.* 2020). Relative to a single excitation and emission wavelength pair, EEM provides more abundant information, which is beneficial to improve the sensitivity of detection (Li *et al.* 2020). Fluorescence regional integration (FRI) is a quantitative method for analyzing EEM, which was first used to characterize fluorescence spectra of dissolved organic matter (DOM) in water (Chen *et al.* 2003). The principle of this method is to divide an EEM contour map into several regions which correspond to different fluorophores and calculate the FI integral of each region. Some studies have reported that FRI was an effective method which could be used to assess the formation potentials of disinfection by-products and the humification degree (Yang *et al.* 2008; He *et al.* 2011). Nevertheless, few studies so far have applied FRI to characterize fluorescence spectra of bacteria.

Consequently, the objective of this study was to analyze the fluorescence characteristics of four different bacterial species and indigenous bacteria in secondary effluent from two water reclamation plants (WRPs) and develop a quantification method on bacterial autofluorescence by FRI, thus providing insight into the rapid detection and quantification of bacteria in water.

2. MATERIALS AND METHODS

2.1. Bacterial samples preparation

Two Gram-positive bacteria, *Bacillus subtilis* (CICC 10275) and *Staphylococcus aureus* (CGMCC 1.12409), and two Gram-negative bacteria, *Escherichia coli* (CGMCC 1.3373) and *Pseudomonas aeruginosa* (CGMCC 1.12483), were used for analysis. For each experiment, bacterial cultures stored in 20% glycerol at −80 °C were inoculated in 20 ml of Luria Bertani (LB) broth and incubated overnight with shaking (150 rpm) at 37 °C. Bacterial cells were harvested by centrifugation (14,400 g) for 5 min at 4 °C and resuspended in 20 ml of 0.9% sterile saline solution. After repeating the washing steps twice, the final bacterial suspensions were 10-fold diluted serially with 0.9% sterile saline solution and used for fluorescence detection. Bacterial concentration was determined by plating 0.1 ml of bacterial solutions in triplicate on LB agar plate and counting colony-forming units (CFU) after overnight incubation at 37 °C.

2.2. Secondary effluent sample preparation

Secondary effluent was sampled from two large-scale WRPs in Beijing, China. Plant A has a daily treatment capacity of 1×10^6 m^3/d and a service area of 97 km^2, which undertakes the wastewater treatment of the downtown area and the eastern area of Beijing. Plant B has a daily treatment capacity of 550,000 m^3/d, which undertakes the wastewater treatment of the western area of Beijing. 200 ml of water samples were filtered through a 0.22 μm sterile filter membrane and bacteria enriched on the membrane were eluted with 20 ml of 0.9% sterile saline solution. 10-fold serial dilutions of bacterial eluents were taken with 0.9% sterile saline solution and used for fluorescence detection. Bacterial concentration was determined by plating 0.1 ml of bacterial solutions in triplicate on an R2A agar plate and counting CFU after incubation for 48 h at 25 °C.

2.3. Fluorescence EEM measurements

Fluorescence EEM measurements were carried out by a fluorescence spectrophotometer (F-7100, Hitachi, Japan). The excitation wavelength (EX) ranged from 200 to 500 nm with 5-nm intervals, and the emission wavelength (EM) ranged from 250 to 600 nm with 2-nm intervals. The excitation and emission slits were 5 nm, and the voltage of photomultiplier was 600 V. Each sample was measured in duplicate, and the scanning time of each measurement was about 2 min. All measurements were finished at room temperature (20 ± 2 °C).

2.4. Data processing and interpretation

Overall, 10,736 data points were obtained for each EEM, and some data points (EM $<$ EX or EM $>$ 2 EX) were removed considering first-order and second-order Rayleigh scattering. Besides, the Raman scattering intensity of 0.9% saline solution was deducted accordingly. The EEM contour map was divided into three regions, and the FI integral of each region was calculated by MATLAB software (R2020b, MathWorks, USA). The correlation analysis between the logarithm of the bacterial concentration and the average logarithm of the FI integral was carried out with Excel software (2019, Microsoft, USA). Principal component analysis (PCA) was applied to differentiate the emission spectra at the excitation wavelength of 280 nm using Origin software (2021, OriginLab, USA).

3. RESULTS AND DISCUSSION

3.1. Fluorescence characteristics of four bacterial species

According to the literature, bacterial intrinsic fluorophores mainly include aromatic amino acids (i.e., tryptophan, tyrosine and phenylalanine) and coenzymes (Ammor 2007; Faassen & Hitzmann 2015). Particularly, tryptophan fluorescence (maximal EX: 280 nm, maximal EM: 350 nm) is the most intense among the three aromatic amino acids, which has been widely used in the detection and quantification of bacteria in water (Baker *et al.* 2015; Sorensen *et al.* 2015). Tyrosine fluorescence (maximal EX: 275 nm, maximal EM: 300 nm) and phenylalanine fluorescence (maximal EX: 260 nm, maximal EM: 280 nm) usually account for less than 10% of the total fluorescence from amino acids (Dalterio *et al.* 1986). Coenzymes in bacteria contain reduced nicotinamide adenine dinucleotide (NADH), reduced NADH phosphate (NADPH) and flavin adenine dinucleotide (FAD). Notably, NADH (maximal EX: 351 nm, maximal EM: 460 nm), NADPH (maximal EX: 336 nm, maximal EM: 464 nm) and FAD (maximal EX: 450 nm, maximal EM: 535 nm) also play an important role in the autofluorescence of bacteria since these coenzymes fluorophores are closely related to bacterial metabolism, which can be used to distinguish bacteria from inanimate materials (Estes *et al.* 2003). Therefore, this study proposed a new FRI approach by dividing bacterial intrinsic fluorophores into three regions, namely Region A (EX: 200–300 nm, EM: 250–400 nm) which is related to amino acids, Region N (EX: 300–400 nm, EM: 400–500 nm) which is related to NAD(P)H and Region F (EX: 400–500 nm, EM: 500–600 nm) which is related to flavins (Figure 1). Compared to five EEM regions of DOM (Chen *et al.* 2003), this new FRI method based on bacterial intrinsic fluorophores enlarged the ranges of EX and EM and added Regions N and F to better reflect the autofluorescence of bacteria, which is beneficial for comprehensive detection of bacterial contamination in water.

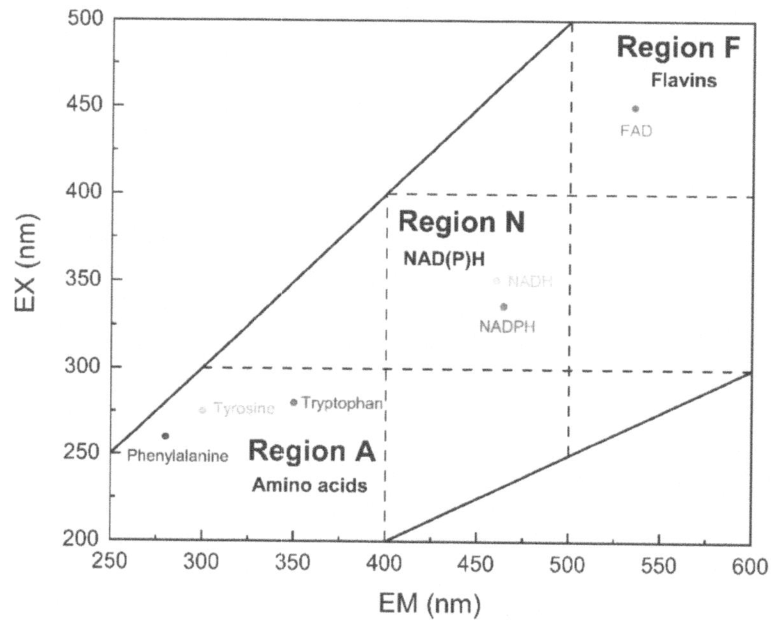

Figure 1 | EEM regions of bacterial intrinsic fluorophores.

Based on the new FRI division depicted in Figure 1, the EEM contours of four typical bacterial species were further measured (Figure 2). The results showed that most of the bacterial autofluorescence was emitted by amino acids (Region A), while coenzymes fluorescence (Regions N and F) was much less intense. This is probably because the main constituents of bacterial cells are proteins, which are made up of amino acids. However, the content of coenzymes in bacteria is relatively low. There were two strong fluorescence peaks in Region A, although their maximal EX was different (around 230 and 280 nm), and the maximal EM was the same (around 330 nm). Dartnell *et al.* (2013) investigated the fluorescence characteristics of several clinically important bacteria and found a double-peak fluorescence associated with tryptophan, which was consistent with the results of this study. To prove this conclusion, the EEM contours of the standard tryptophan were measured, and a double-peak fluorescence was also observed (Supplementary Figure S1). Although the four bacterial species all exhibited tryptophan fluorescence in Region A, the FI of different bacterial species was different. For example, the FI values of *B. subtilis* and *S. aureus* in Region A were higher than those of *E. coli* and *P. aeruginosa*, indicating that *B. subtilis* and *S. aureus* were more abundant in proteins. This may be due to the difference in cell wall structure between Gram-positive and Gram-negative bacteria. Especially, although the concentration of *B. subtilis* was much lower than that of the other three bacterial species, the FI of *B. subtilis* was comparable among the four bacterial species, which could be attributed to the content of extracellular polymeric substance (EPS) excreted by bacteria since proteins in EPS also exhibited fluorescence (Marvasi *et al.* 2010). Moreover, *P. aeruginosa* also showed strong fluorescence in Region N due to the siderophore

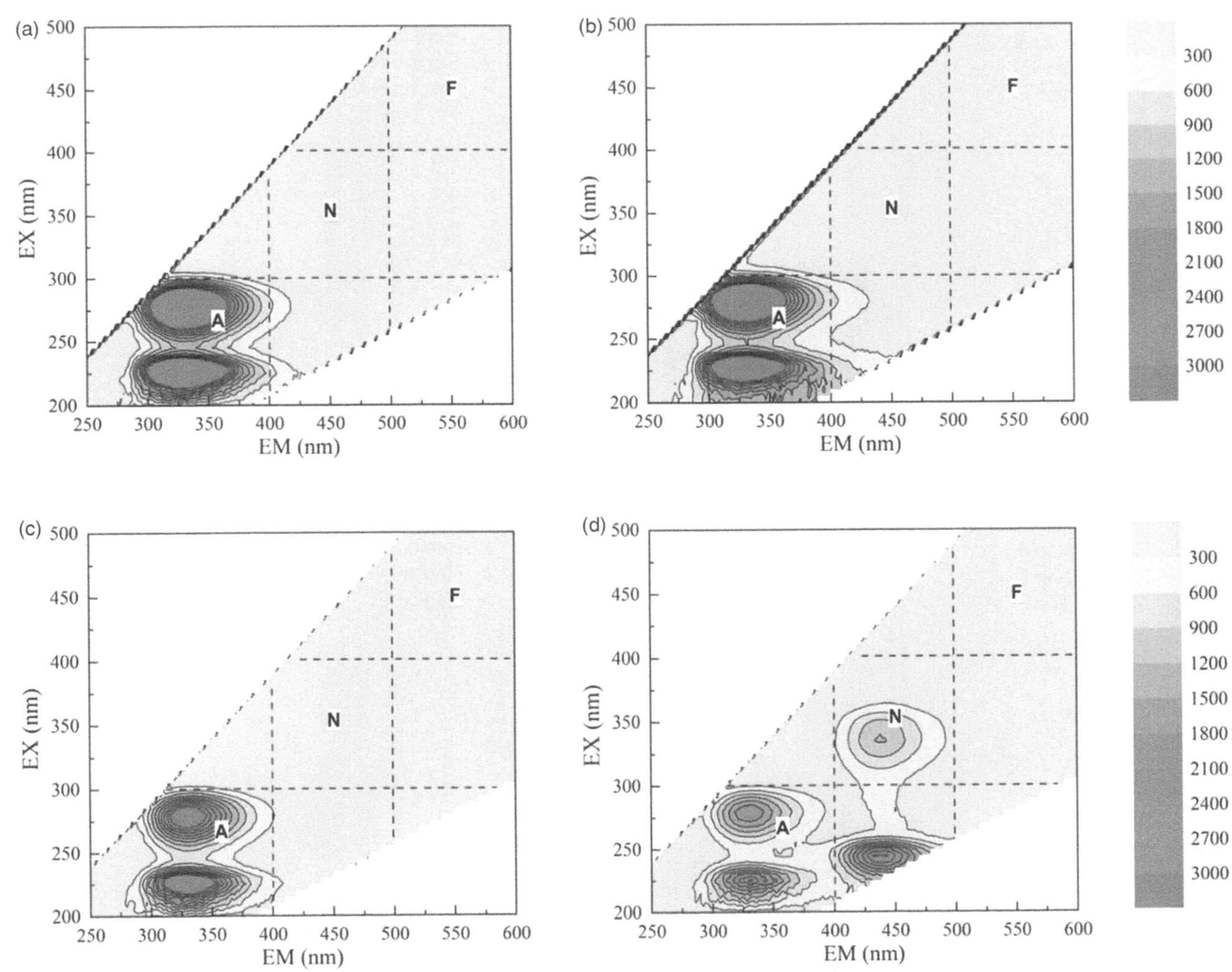

Figure 2 | EEM contours of (a) *B. subtilis* (8×10^4 CFU/ml), (b) *S. aureus* (1.65×10^7 CFU/ml), (c) *E. coli* (4.3×10^7 CFU/ml), and (d) *P. aeruginosa* (2.75×10^7 CFU/ml).

pyoverdine in EPS (Cornelis 2010). Overall, all the four bacterial species exhibited a double-peak fluorescence in Region A, but the FI varied largely among different bacterial species.

3.2. Correlation analysis between FI integral and bacterial concentration

The relationship between FI integral and bacterial concentration was further analyzed, as shown in Figure 3. With the increase of bacterial concentration, the FI integral of each region also increased. The FI integral of Region A was one or two orders of magnitude higher than that of Regions N and F except for *P. aeruginosa*, which was consistent with EEM contours. Table 1 lists the determination coefficient of linear regression analysis (R^2) and slope between the logarithm of FI integral and the logarithm of bacterial concentration. For each bacterial species, R^2 of Region F were all above 0.9, indicating that the FI integral of Region F had a good linear relationship with bacterial concentration. Although tryptophan fluorescence is widely used in the quantification of bacteria, the results from this study suggested that flavin fluorescence could offer a better indication of bacterial concentration. Furthermore, tryptophan fluorescence is not suitable for characterizing bacteria in the case of high organic contents in actual water and wastewater samples since organic matters such as proteins also exhibit fluorescence in Region A, which makes it difficult for distinguishing between organic matters and bacteria. However, the detection limit based on the FI integral of Region F varied among the four bacterial species in this study. For example, the detection limit of *B. subtilis* was as low as 10 CFU/ml, while *E. coli* could be detected only if the concentration was above 10^6 CFU/ml. This is because the content of flavins in bacteria depends on the type and metabolic state of the bacteria

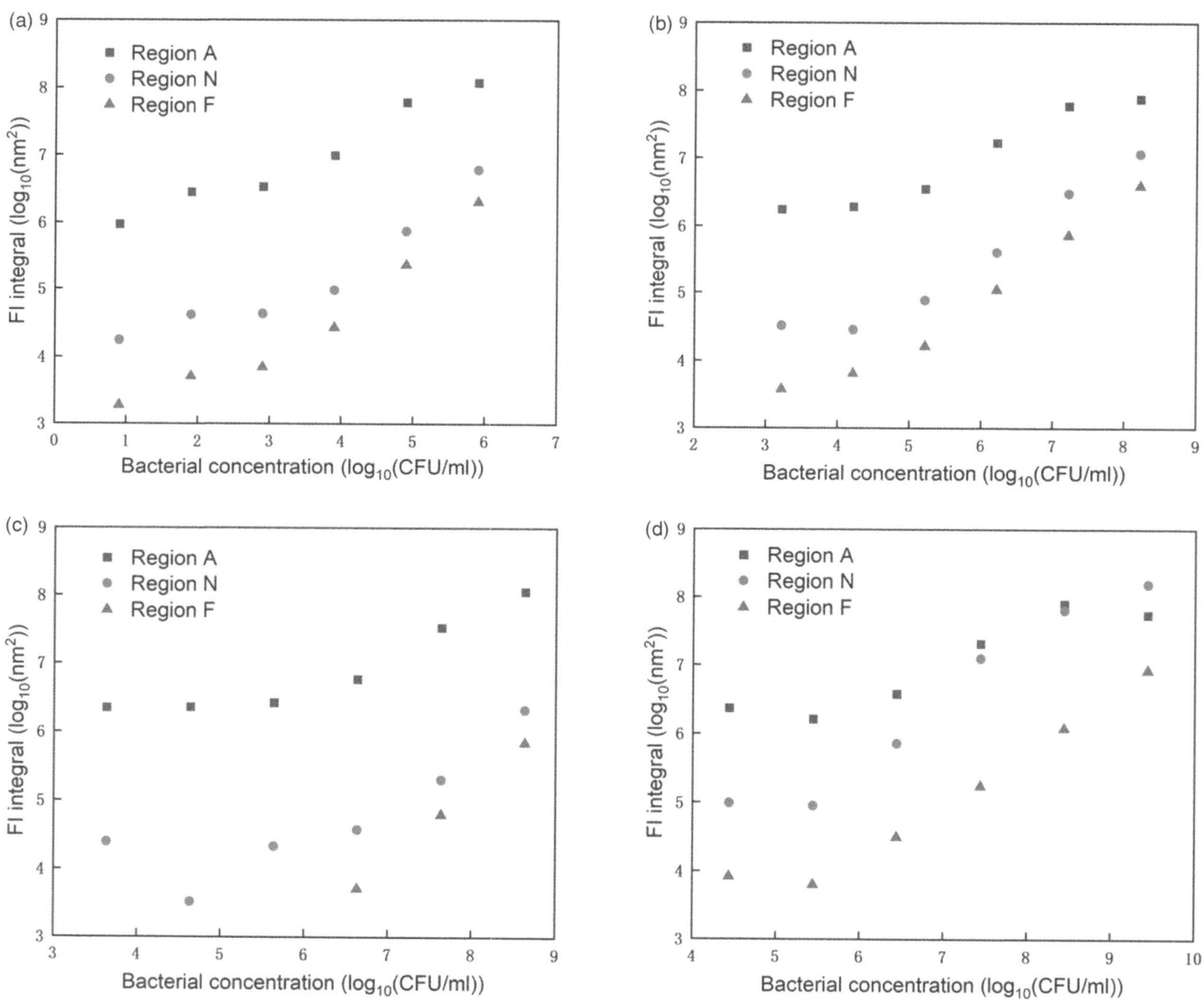

Figure 3 | Relationship between FI integral and bacterial concentration: (a) *B. subtilis*, (b) *S. aureus*, (c) *E. coli*, and (d) *P. aeruginosa*.

Table 1 | Determination coefficient of linear regression analysis (R^2) and slope between the logarithm of FI integral and the logarithm of bacterial concentration

Region	Statistics	*B. subtilis*	*S. aureus*	*E. coli*	*P. aeruginosa*
A	R^2	0.9579	0.9352	0.8527	0.8642
	Slope	0.43	0.38	0.35	0.36
N	R^2	0.8813	0.9323	0.7175	0.9504
	Slope	0.48	0.56	0.44	0.74
F	R^2	0.9330	0.9654	0.9999	0.9420
	Slope	0.59	0.63	1.06	0.65

and coenzyme fluorescence is not always detectable (Estes *et al.* 2003). Therefore, although flavin fluorescence was well correlated with bacterial concentration, it is still needed to strengthen its fluorescence response and lower the detection limit.

3.3. Fluorescence detection of bacteria in secondary effluents

To explore the feasibility of applying fluorescence spectroscopy to bacterial detection in actual water samples, the EEM contours of indigenous bacteria in secondary effluent from two WRPs were measured (Figure 4). Similar to pure bacteria, amino acids accounted for most of the bacterial fluorescence and a double-peak fluorescence could also be observed in Region A. Although the concentration of bacteria from Plants A and B was on the same order of magnitude, the FI of bacteria from Plant A was much higher than that from Plant B. This is probably because of the differences in bacterial community structure between the two plants (Supplementary Figure S7). Although the biological treatment process of the two plants is the same (i.e., Anaerobic–Anoxic–Oxic, A2/O), the source of wastewater is different, which results in different dominant bacteria in water. Plant A receives wastewater mainly from the downtown area, while Plant B receives wastewater from the western area of Beijing. Figure 5 shows the relationship between FI integral and bacterial concentration. Table 2 lists the determination coefficient of linear regression analysis (R^2) and slope between the logarithm of FI integral and the logarithm of bacterial concentration. As for Plant A, there was a good linear correlation ($R^2 > 0.9$) between the FI integral of three regions and bacterial concentration. However, as for Plant B, the FI integral of Region A had a poor correlation ($R^2 = 0.6597$) with bacterial concentration, and the detection limit based on Region N was above 10^3 CFU/ml. Flavin fluorescence seemed to correlate well with bacterial concentration in both plants. Therefore, the double-peak tryptophan fluorescence in Region A can be used to characterize the existence of bacteria, while the FI integral of Region F can be used to characterize bacterial concentration in secondary effluent. It is noted that the bacterial community structure is different in different types of water, and thus even with the same bacterial concentration, the FI may vary greatly. The quantification method based on

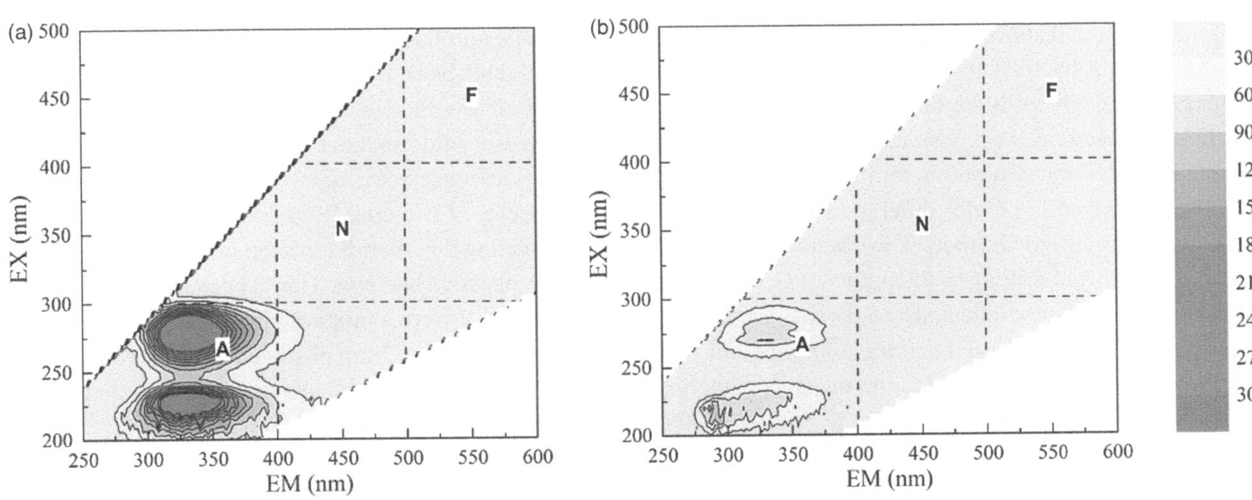

Figure 4 | EEM contours of bacteria in secondary effluent from (a) Plant A (6.2×10^5 CFU/ml) and (b) Plant B (1.6×10^5 CFU/ml).

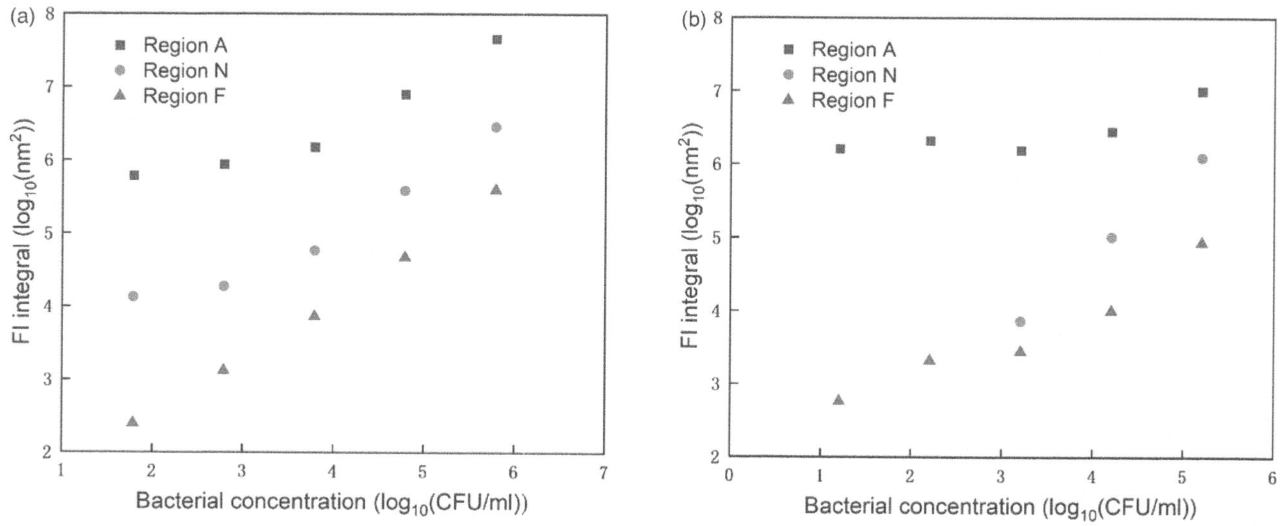

Figure 5 | Relationship between FI integral and bacterial concentration: (a) Plant A and (b) Plant B.

Table 2 | Determination coefficient of linear regression analysis (R^2) and slope between the logarithm of FI integral and the logarithm of bacterial concentration

Region	Statistics	Plant A	Plant B
A	R^2	0.9117	0.6597
	Slope	0.47	0.17
N	R^2	0.9366	0.9997
	Slope	0.60	1.11
F	R^2	0.9975	0.9335
	Slope	0.79	0.50

FRI can be used for long-term monitoring of a certain type of water, which has an early warning effect to bacterial concentration exceeding standard limits and/or potential microbial risks.

3.4. PCA for differentiation of bacteria

PCA was applied to the emission spectra (EX: 280 nm) of these four bacterial species and indigenous bacteria from two WRPs (Figure 6). As mentioned above, tryptophan fluorescence is the characteristic autofluorescence of bacteria. However, the four bacterial species and bacteria from secondary effluents all exhibited a double-peak tryptophan fluorescence in Region A, which makes it difficult to differentiate them. To explore the possibility of whether fluorescence spectroscopy could distinguish different bacteria, PCA was used to extract feature information from fluorescence spectra. In the PCA diagram, the distance between any two points represents the similarity of their fluorescence spectra. Particularly, PC1 (64.69%) and PC2 (33.20%) explained the major difference among the fluorescence spectra of bacteria. It was observed that the distance between Gram-positive and Gram-negative bacteria was relatively far, indicating that their fluorescence spectra varied greatly. The points of Plants A and B lay in the points of Gram-positive and Gram-negative bacteria. This is because the components in secondary effluent are complicated and diverse, bacteria from Plants A and B were a mixture of varied Gram-positive and Gram-negative bacteria. It is noted that the point of Plant A was closer to the point of Gram-positive bacteria, while the point of Plant B was closer to the point of Gram-negative bacteria, which was consistent with the above conclusion that the FI of the former was higher than that of the latter. Sohn *et al.* (2009) also used PCA to analyze the emission spectra (EX: 280 nm) of *E. coli*, *Salmonella* and *Campylobacter* and found that different bacterial species as well as their concentrations could be successfully classified. According to the PCA results from this study, different bacterial species could be distinguished, and the composition of bacterial community structure in secondary effluent from different WRPs could be roughly estimated

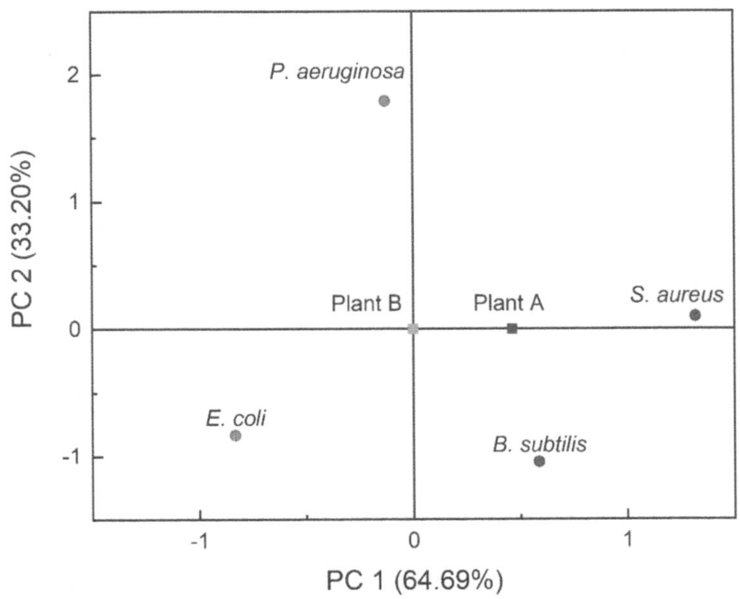

Figure 6 | PCA of the emission spectra (EX: 280 nm) of *B. subtilis* (8×10^4 CFU/ml), *S. aureus* (1.65×10^7 CFU/ml), *E. coli* (4.3×10^7 CFU/ml), *P. aeruginosa* (2.75×10^7 CFU/ml), and bacteria from Plant A (6.2×10^5 CFU/ml) and Plant B (1.6×10^5 CFU/ml).

according to the distance between Gram-positive and Gram-negative bacteria. Thus, fluorescence spectroscopy coupled with PCA analysis will be helpful for the rapid determination of microbial water quality fluctuations.

3.5. Discussion

3.5.1. Possible applications

Fluorescence spectroscopy has a good prospect in the rapid detection of bacteria in water, and there are three possible applications. First, real-time online monitoring of bacterial concentration in bacteria-rich water samples (e.g., wastewater and reclaimed water) can be achieved by automatic measurement of fluorescence EEM of water samples after filtration. With the relationship between the FI integral of Region F and bacterial concentration, the quantitative detection of bacteria in water samples can be achieved by automatically calculating the FI integral of water samples. PCA can be further used to determine whether bacterial community structure has changed or not. Traditional culture-based methods are time-consuming and labor-intensive, and accidental hazardous events may have happened before the results are obtained. Therefore, the serious lag of culture-based methods makes them unsuitable for real-time monitoring. Comparatively, the measurement of fluorescence EEM only takes 2–3 min without a high level of expertise compared to PCR and FCM techniques, which also reduces the generation of human errors. Secondly, fluorescence spectroscopy can be used to detect bacterial contamination in relatively clean water (e.g., drinking water). According to the results in this study, once the highly purified water is contaminated with bacteria, substantial changes of fluorescence signals in Region A can be observed. Therefore, in addition to the quantitative monitoring of bacterial concentration, the qualitative detection of bacterial contamination is also feasible. Thirdly, a portable spectrometer based on the detection of fluorescence peaks can be exploited for more field investigation on water quality. In fact, fluorescence spectroscopy can be used for the characterization of both organic matters and microorganisms (Bridgeman *et al.* 2015). Although the results by portable spectrometer may not be as accurate as fluorescence EEM, portability and faster detection time make it superior when it comes to early warning of water contamination.

3.5.2. Current limitations and future research directions

Although fluorescence spectroscopy has more advantages in the detection of bacteria in water compared to other methods, there are some limitations. For one thing, the content of intrinsic fluorophores, especially for coenzymes in bacteria, is relatively low. Thus, in this study, the detection limit based on the FI integral of Regions N and F varied greatly among different bacterial species. In other words, the FRI method may not be sensitive enough when it comes to low bacterial concentration, which makes it difficult for the quantification of bacteria in drinking water. Although bacteria can be enriched through filtration, the additional step may also increase the detection error. Therefore, enhancing the coenzyme fluorescence in bacteria can be further explored for quantitative detection when it comes to low bacterial concentration (Chowdhury *et al.* 2011). For another, organic matters (e.g., proteins) in water also emit fluorescence, and sometimes the FI is much higher than the bacterial intrinsic autofluorescence, which may cause interference or even masking to bacterial detection (Sherchan *et al.* 2018). Although fluorescence spectroscopy can be used to quantify bacterial concentration in relatively polluted water, fluorescence signals of organic matters under high DOM concentration could not be ignored. Further work is required to distinguish bacteria from organic matters. For example, the fluorescence data processing can be used to correct the fluorescence internal filtration effect and remove the redundant data to find out the characteristic information of bacteria (Sgroi *et al.* 2020). In addition, parallel factor analysis (PARAFAC) is also a data dimension reduction method which has been increasingly used for the identification of various fluorophores (Murphy *et al.* 2013). Fluorescence spectroscopy coupled with FRI, PCA and PARAFAC can be a research area of concern in the field of the rapid detection and quantification of bacteria in water.

4. CONCLUSIONS

In this study, a quantitative method for bacteria by EEM FRI was developed, and bacterial intrinsic fluorophores were divided into three regions, including Region A (amino acids), Region N (NAD(P)H) and Region F (flavins). The autofluorescence of four bacterial species (*B. subtilis*, *S. aureus*, *E. coli* and *P. aeruginosa*) was identified and mainly concentrated in Region A, implying that the main bacterial intrinsic fluorophores were amino acids. It was found that the FI integral of Region F had a good linear relationship ($R^2 > 0.9$) with bacterial concentration. In addition, the study demonstrated that PCA could distinguish varied bacterial species and bacteria from different secondary effluents. In conclusion, fluorescence spectroscopy shows great potential in the rapid detection of bacteria in water, and further work is still needed to improve the sensitivity and applicability of the FRI method.

ACKNOWLEDGEMENTS

This study was supported by the National Key Research R&D Program of China (No. 2016YFC0401405) and the Key Program of the National Natural Science Foundation of China (No. 51738005).

DATA AVAILABILITY STATEMENT

All relevant data are included in the paper or its Supplementary Information.

REFERENCES

Ammor, M. S. 2007 Recent advances in the use of intrinsic fluorescence for bacterial identification and characterization. *Journal of Fluorescence* **17** (5), 455–459.

Baker, A., Cumberland, S. A., Bradley, C., Buckley, C. & Bridgeman, J. 2015 To what extent can portable fluorescence spectroscopy be used in the real-time assessment of microbial water quality? *Science of the Total Environment* **532**, 14–19.

Banihashemi, A., Van Dyke, M. I. & Huck, P. M. 2015 Detection of viable bacterial pathogens in a drinking water source using propidium monoazide-quantitative PCR. *Journal of Water Supply Research and Technology – AQUA* **64** (2), 139–148.

Bottari, B., Santarelli, M. & Neviani, E. 2015 Determination of microbial load for different beverages and foodstuff by assessment of intracellular ATP. *Trends in Food Science & Technology* **44** (1), 36–48.

Bridgeman, J., Baker, A., Brown, D. & Boxall, J. B. 2015 Portable LED fluorescence instrumentation for the rapid assessment of potable water quality. *Science of the Total Environment* **524**, 338–346.

Budnik, L. T. & Casteleyn, L. 2019 Mercury pollution in modern times and its socio-medical consequences. *Science of the Total Environment* **654**, 720–734.

Cangelosi, G. A. & Meschke, J. S. 2014 Dead or alive: Molecular assessment of microbial viability. *Applied and Environmental Microbiology* **80** (19), 5884–5891.

Carstea, E. M., Bridgeman, J., Baker, A. & Reynolds, D. M. 2016 Fluorescence spectroscopy for wastewater monitoring: a review. *Water Research* **95**, 205–219.

Chen, W., Westerhoff, P., Leenheer, J. A. & Booksh, K. 2003 Fluorescence excitation–emission matrix regional integration to quantify spectra for dissolved organic matter. *Environmental Science & Technology* **37** (24), 5701–5710.

Chen, Z., Wu, G., Wu, Y., Wu, Q. & Hu, H. Y. 2020 Water Eco-Nexus Cycle System (WaterEcoNet) as a key solution for water shortage and water environment problems in urban areas. *Water Cycle* **1**, 71–77.

Chowdhury, M. H., Lakowicz, J. R. & Ray, K. 2011 Ensemble and single molecule studies on the use of metallic nanostructures to enhance the intrinsic emission of enzyme cofactors. *Journal of Physical Chemistry C* **115** (15), 7298–7308.

Clemens, J. D., Nair, G. B., Ahmed, T., Qadri, F. & Holmgren, J. 2017 Cholera. *Lancet* **390** (10101), 1539–1549.

Cornelis, P. 2010 Iron uptake and metabolism in pseudomonads. *Applied Microbiology and Biotechnology* **86** (6), 1637–1645.

Dalterio, R. A., Nelson, W. H., Britt, D., Sperry, J., Psaras, D., Tanguay, J. F. & Suib, S. L. 1986 Steady-state and decay characteristics of protein tryptophan fluorescence from bacteria. *Applied Spectroscopy* **40** (1), 86–90.

Dartnell, L. R., Roberts, T. A., Moore, G., Ward, J. M. & Muller, J.-P. 2013 Fluorescence characterization of clinically-important bacteria. *Plos One* **8** (9), e75270.

Enning, D. & Garrelfs, J. 2014 Corrosion of iron by sulfate-reducing bacteria: new views of an old problem. *Applied and Environmental Microbiology* **80** (4), 1226–1236.

Estes, C., Duncan, A., Wade, B., Lloyd, C., Ellis, W. & Powers, L. 2003 Reagentless detection of microorganisms by intrinsic fluorescence. *Biosensors & Bioelectronics* **18** (5–6), 511–519.

Faassen, S. M. & Hitzmann, B. 2015 Fluorescence spectroscopy and chemometric modeling for bioprocess monitoring. *Sensors* **15** (5), 10271–10291.

Giana, H. E., Silveira, L., Zangaro, R. A. & Pacheco, M. T. T. 2003 Rapid identification of bacterial species by fluorescence spectroscopy and classification through principal components analysis. *Journal of Fluorescence* **13** (6), 489–493.

He, X.-S., Xi, B.-D., Wei, Z.-M., Jiang, Y.-H., Yang, Y., An, D., Cao, J.-L. & Liu, H.-L. 2011 Fluorescence excitation-emission matrix spectroscopy with regional integration analysis for characterizing composition and transformation of dissolved organic matter in landfill leachates. *Journal of Hazardous Materials* **190** (1–3), 293–299.

Li, L., Wang, Y., Zhang, W. J., Yu, S. L., Wang, X. Y. & Gao, N. Y. 2020 New advances in fluorescence excitation-emission matrix spectroscopy for the characterization of dissolved organic matter in drinking water treatment: a review. *Chemical Engineering Journal* **381**, 122676.

Luo, L. W., Wu, Y. H., Wang, Y. H., Tong, X., Bai, Y., Chen, G. Q., Wang, H. B., Ikuno, N. & Hu, H. Y. 2021 Aggravated biofouling caused by chlorine disinfection in a pilot-scale reverse osmosis treatment system of municipal wastewater. *Journal of Water Reuse and Desalination* **11** (2), 201–211.

Marvasi, M., Visscher, P. T. & Martinez, L. C. 2010 Exopolymeric substances (EPS) from *Bacillus subtilis*: polymers and genes encoding their synthesis. *Fems Microbiology Letters* **313** (1), 1–9.

Matin, A., Khan, Z., Zaidi, S. M. J. & Boyce, M. C. 2011 Biofouling in reverse osmosis membranes for seawater desalination: phenomena and prevention. *Desalination* **281**, 1–16.

Murphy, K. R., Stedmon, C. A., Graeber, D. & Bro, R. 2013 Fluorescence spectroscopy and multi-way techniques. PARAFAC. *Analytical Methods* **5** (23), 6557–6566.

Nakar, A., Schmilovitch, Z. E., Vaizel-Ohayon, D., Kroupitski, Y., Borisover, M. & Sela, S. 2020 Quantification of bacteria in water using PLS analysis of emission spectra of fluorescence and excitation-emission matrices. *Water Research* **169**, 115197.

Ntziachristos, V., Ripoll, J., Wang, L. H. V. & Weissleder, R. 2005 Looking and listening to light: the evolution of whole-body photonic imaging. *Nature Biotechnology* **23** (3), 313–320.

Rajapaksha, P., Elbourne, A., Gangadoo, S., Brown, R. C. D., Cozzolino, D. & Chapman, J. 2019 A review of methods for the detection of pathogenic microorganisms. *Analyst* **144** (2), 396–411.

Rollins, D. M. & Colwell, R. R. 1986 Viable but nonculturable stage of *Campylobacter jejuni* and its role in survival in the natural aquatic environment. *Applied and Environmental Microbiology* **52** (3), 531–538.

Safford, H. R. & Bischel, H. N. 2019 Flow cytometry applications in water treatment, distribution, and reuse: a review. *Water Research* **151**, 110–133.

Sgroi, M., Gagliano, E., Vagliasindi, F. G. A. & Roccaro, P. 2020 Inner filter effect, suspended solids and nitrite/nitrate interferences in fluorescence measurements of wastewater organic matter. *Science of the Total Environment* **711**, 134663.

Sherchan, S. P., Miles, S. L., Ikner, L. A., Yu, H., Snyder, S. A. & Pepper, I. L. 2018 Near real-time detection of *E. coli* in reclaimed water. *Sensors* **18** (7), 2303.

Shi, X., Chen, Z., Lu, Y., Shi, Q., Wu, Y. & Hu, H.-Y. 2020 Significant increase of assimilable organic carbon (AOC) levels in MBR effluents followed by coagulation, ozonation and combined treatments: implications for biostability control of reclaimed water. *Frontiers of Environmental Science & Engineering* **15** (4), 68.

Sohn, M., Himmelsbach, D. S., Barton II, F. E. & Fedorka-Cray, P. J. 2009 Fluorescence spectroscopy for rapid detection and classification of bacterial pathogens. *Applied Spectroscopy* **63** (11), 1251–1255.

Sorensen, J. P. R., Lapworth, D. J., Marchant, B. P., Nkhuwa, D. C. W., Pedley, S., Stuart, M. E., Bell, R. A., Chirwa, M., Kabika, J., Liemisa, M. & Chibesa, M. 2015 In-situ tryptophan-like fluorescence: a real-time indicator of faecal contamination in drinking water supplies. *Water Research* **81**, 38–46.

UN 2015 *The Sustainable Development Goals*. United Nations, New York.

WHO 2011 *Guidelines for Drinking-water Quality*, 4th edn. World Health Organization, Geneva.

Yang, X., Shang, C., Lee, W., Westerhoff, P. & Fan, C. 2008 Correlations between organic matter properties and DBP formation during chloramination. *Water Research* **42** (8–9), 2329–2339.

First received 26 May 2021; accepted in revised form 15 July 2021. Available online 6 August 2021

Water Recycling and Low-Carbon Development doi: 10.2166/wrd.2021.043

Effects of mixed irrigation using brackish water with different salinities and reclaimed water on a soil-crop system

Chuncheng Liu [a,b,c], Bingjian Cui [a,c], Chao Hu [a,c], Haiqing Wu [a,c] and Feng Gao [a,c,*]

[a] Institute of Farmland and Irrigation, CAAS/Key Lab of Water-saving Irrigation Engineering, Ministry of Agriculture, Xinxiang 453002, China
[b] School of Chinese Academy of Agricultural Sciences, Beijing 100081, China
[c] Agriculture Water and Soil Environmental Field Science Research Station of Xinxiang City of Henan Province of CAAS, Xinxiang 453000, China
*Corresponding author. E-mail: gaofengcaas@sina.com

CL, 0000-0001-9769-0009; FG, 0000-0003-1590-2325

ABSTRACT

To explore the effects of mixed irrigation on soil and crops, a pot experiment was conducted in two salinity levels of brackish water, four levels of mixed brackish-reclaimed water and freshwater irrigation as the control. The soil Na-Cl to Ca-SO_4 contents changed, and activities of soil alkaline phosphatase and polyphenol oxidase changed, exhibiting a 'V'-shaped curve with increasing the proportion of reclaimed water in the mixture. At the same brackish-reclaimed water level, there were no significant differences in alkaline phosphatase and polyphenol oxidase activities except for soil alkaline phosphatase activity decreasing significantly with the increase in salinity under brackish water irrigation. Mixed irrigation obviously improved superoxide dismutase activity but no significant influences on aboveground dry weight, underground biomass or crop physiological indexes (chlorophyll, soluble protein, malondialdehyde, peroxidase, catalase). Based on the integrated biological response index version 2 (IBRv2), the deviation of reclaimed water irrigation was the smallest, followed by 1:1 and 1:2 (3, 5 g/L brackish water salinities, respectively), with IBRv2 values of 7.94, 12.55 and 16.04. Therefore, considering the soil-crop characteristics, limited daily water amount and inadequate pipeline facilities for reclaimed water, the brackish-reclaimed water ratio should be 1:1 and 1:2 at 3, 5 g/L of brackish water, respectively.

Key words: brackish water, crop, IBRv2, mixed irrigation, reclaimed water, soil

HIGHLIGHTS

- Mixed irrigation with brackish water-reclaimed water was introduced.
- IBRv2 was used to evaluate the effects of mixed irrigation on soil-crop system.
- Brackish-reclaimed water ratio should be 1:1 and 1:2 at 3 g/L and 5 g/L of irrigation water salinity, respectively.

GRAPHICAL ABSTRACT

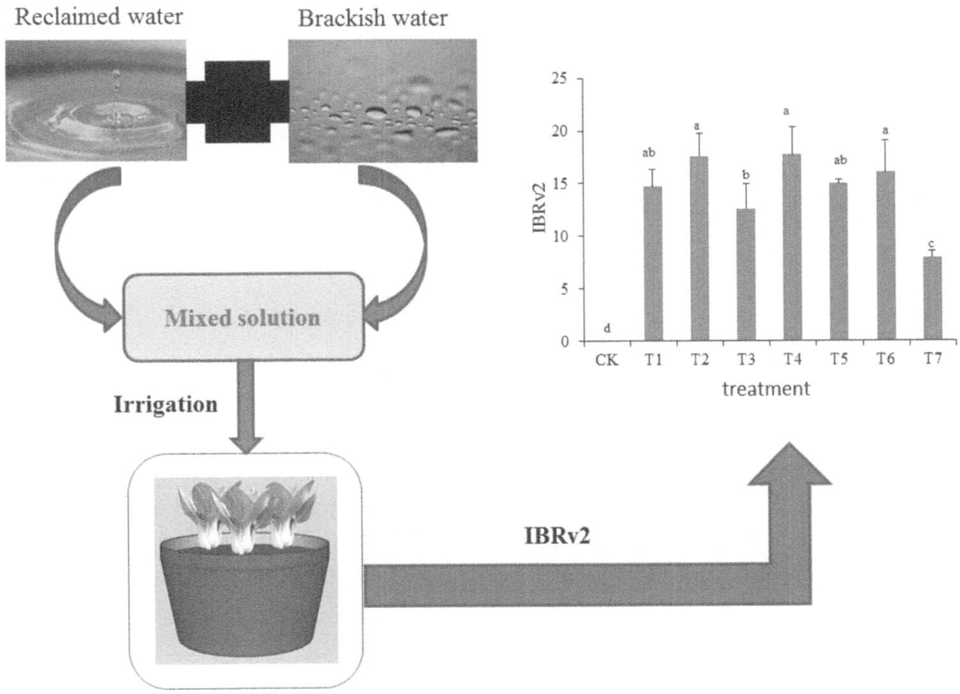

1. INTRODUCTION

China has water sources, but they are unevenly spread throughout the country. China's agricultural water consumption accounts for more than 60% of the total water consumption, but the mismatch between the temporal-spatial distributions of water resources and cultivated land resources makes it difficult to meet the demand for agricultural water in northern China, which seriously restricts the sustainable development of agriculture. China is rich in unconventional water resources such as reclaimed water and brackish water. In China, the irrigation amounts of reclaimed water and brackish water in 2015 reached 11.01 billion and 1.48 billion m^3, respectively (Hu & Wu 2018). Therefore, it is important to use unconventional water resources for irrigation to relieve the pressure of insufficient freshwater resources.

In recent years, research on brackish water utilization has been carried out mainly from the aspects of brackish water salinity, soil texture, suitable crops and field management, and a relatively complete technical system has gradually formed. In most salinized agricultural areas lacking freshwater resources, brackish water irrigation has become an important way to improve saline-alkali land and increase crop yield (Huang *et al.* 2019). Soil salinity affects the water retention of silty clay mainly in the high-suction range and decreases the plant-available water capacity (Tang *et al.* 2021). The water electrical conductivities linked to irrigation levels have a direct effect on the morphophysiological characteristics of collards (Viana *et al.* 2021). Within a certain range, brackish water irrigation can stimulate crop growth without a significant reduction or increase in crop yield (United States Salinity Laboratory Staff 1954; Yuan *et al.* 2019) and can improve water use efficiency (Yuan *et al.* 2019). Short-term brackish water irrigation has no obvious effect on soil chemical properties and soil salinization, while long-term brackish water irrigation may cause soil salinization (Tahtouh *et al.* 2019) and affect crop growth (Feng *et al.* 2014). However, some researchers believe that brackish water irrigation can increase the grain protein content (Cucci *et al.* 2019). In addition, brackish water irrigation may induce the occurrence of soil water repellency (Liu *et al.* 2011). Freshwater irrigation (1.2 dS/m) in the vegetative growth stage can improve the potential yield, and saline water irrigation (7 dS/m) in the reproductive stage can improve fruit quality (Bustan *et al.* 2005). HYDRUS simulation results show that long-term saline water (3 g/L) used to irrigate wheat and corn is more suitable for homogeneous soil than heterogeneous soil in North China (Liu *et al.* 2019).

In addition, many studies have been conducted on the safe utilization of reclaimed water. Studies on the utilization of reclaimed water mainly involve the effects of reclaimed water irrigation on crop growth (Bhattacharyya *et al.* 2008;

Han *et al.* 2018a; Perulli *et al.* 2019; Petousi *et al.* 2019) and quality (Sun *et al.* 2007; Hu *et al.* 2013; Pedrero *et al.* 2018), the soil environment (Durán-Alvarez *et al.* 2009; Gibson *et al.* 2010; Han *et al.* 2018a), soil microbial community structure (Guo *et al.* 2017), groundwater (Siemens *et al.* 2008; Lesser *et al.* 2018) and the appropriate irrigation technology of reclaimed water (Hassanli *et al.* 2008; Lu *et al.* 2016; Intriago *et al.* 2018).

However, at present, researchers are mainly focused on brackish water irrigation or reclaimed water irrigation. There are few reports on the related studies of combined irrigation. Therefore, this study set different mixing ratios of brackish water to reclaimed water in pot experiments to explore the effect of mixed irrigation on the soil environment and physiological characteristics of crops to provide a theoretical basis for the safe utilization of reclaimed water in areas irrigated with brackish water.

2. MATERIALS AND METHODS

2.1. Tested soil description

The tested soil was collected from 0 to 20 cm topsoil of a field at Qiliying Experimental Base, Xinxiang city, Henan Province. The soil was air-dried, crushed and sieved (2 mm). The bulk density of the soil was 1.40 g/cm^3, the field water-holding capacity of the soil was 23.02%, total nitrogen and total phosphorus contents in soil were 0.99 and 1.11 g/kg, and the alkali hydrolyzed nitrogen, available phosphorus and available potassium were 90.98, 26.82 and 208.60 mg/kg, respectively (Li *et al.* 2019). The electrical conductivity of a 1:5 soil-water extract was 372 μS/cm, and the mass fraction of organic matter was 2.66%. The particle size distribution of the soil samples was analysed by a BT-9300HT laser particle analyser. The proportions of clay (<0.002 mm), silt (0.002–0.02 mm) and sand (0.02–2 mm) were 13.05%, 62.46% and 24.49%, respectively, and the soil had a silty loam texture (International System).

2.2. Experimental device and scheme

The pot experiment was performed in the greenhouses of the Agriculture Water and Soil Environmental Field Science Research Station at the Chinese Academy of Agricultural Sciences in Xinxiang city, Henan Province, from May to June 2020. The station is located at 35°19′ N, 113°53′ E, 73.2 m above sea level, with an average annual temperature of 14.1 °C and multiyear average annual precipitation and evaporation of 588 mm and 2,000 mm, respectively. The frost-free period lasts for 210 days, and the average annual sunshine duration is 2,398 hr.

The pots had an upper diameter of 25 cm, lower diameter of 14.5 cm and a height of 19 cm. Each pot was loaded with 7 kg of soil, and all treatments received compound fertilizer (the ratio of N-P_2O_5-K_2O was 15-15-15). All of the treatments received fertilizer as a basal application, following the local conventional fertilizer application. The tested crop was Shanghai green, and all treatments were irrigated with freshwater before sowing to maintain moisture. Seeds were sown on 27 May 2020 and were spread evenly in each pot. Five seedlings were left in each pot at the two-leaf stage (11 June), and then the irrigation treatment was started. In the early stage, irrigation was carried out approximately once every 2 days with 400 mL of water (the lower limit of irrigation was 75% of the field capacity); in the later stage, irrigation was performed approximately once per day (400 mL) as the crop's water demand increased. In the experiment, the two salinity levels of the brackish water were set for 3 and 5 g/L, and four mixed ratio levels were set for the mixing ratio of brackish water and reclaimed water, namely, reclaimed water, brackish water-reclaimed water 1:2, brackish water-reclaimed water 1:1, and brackish water. The control group consisted of cultivated crops under freshwater irrigation. The pot was randomly laid, and there were three replicates for each treatment. The specific experimental design is shown in Table 1. The water quality of the brackish water and reclaimed water is shown in Table 2. The reclaimed water was obtained from the Luotuowan Domestic Sewage Treatment Plant in Xinxiang city, Henan Province, and the sewage treatment plant used the A/O process. The water quality after sewage treatment was in line with the Farmland Irrigation Water Quality Standard (GB5084-2005). The freshwater source was local groundwater, and brackish water was prepared by adding sea salt to the freshwater.

Table 1 | Experimental design

Treatment	CK	T1	T2	T3	T4	T5	T6	T7
Solution	FW	3 g/L of BW	5 g/L of BW	1:1 of BW(3 g/L) to RW	1:1 of BW(5 g/L) to RW	1:2 of BW(3 g/L) to RW	1:2 of BW(5 g/L) to RW	RW

FW, freshwater; BW, brackish water; RW, reclaimed water.

Table 2 | Quality of reclaimed water, brackish water and fresh water for experiment

Water source	$EC_{1:5}$/(μS·cm^{-1})	SAR	Ion content/(mg·L^{-1})							
			Na^+	K^+	Ca^{2+}	Mg^{2+}	Cl^-	HCO_3^-	SO_4^{2-}	CO_3^{2-}
Fresh water	321	0.34	10	1.6	39	15	30	120	104	–
Reclaimed water	2,120	5.82	310	13.9	91	74	314	278	507	–
Brackish water (3 g/L)	6,100	43.30	1,330	2.1	43	17	1,921	142	92	–
Brackish water (5 g/L)	9,432	67.19	2,000	2.6	37	18	3,222	139	109	–

2.3. Measured indexes and methods

Measured indexes and methods were as follows:

1. Soil physical and chemical properties. After crop harvesting, soil samples were collected from the pots and were air-dried, ground and passed through a 2-mm sieve. The soil water content was determined by the oven drying method. The conductivity of a 1:5 soil-to-water extract ($EC_{1:5}$) was determined by a conductivity meter. The low-temperature external-heat potassium dichromate oxidation-photocolourimetric method was used to determine the soil organic matter (SOM) content. The 1:5 soil-to-water extracts were prepared, the Na^+ and K^+ contents were determined by flame photometry, the Ca^{2+} and Mg^{2+} contents were determined by EDTA titration, the Cl^- content was determined by $AgNO_3$ titration, the CO_3^{2-} and HCO_3^- contents were determined by double indicator-neutralization titration, and SO_4^{2-} was determined by EDTA indirect complexometric titration. The water drop penetration time (WDPT) method was used to determine the soil WDPT.
2. Soil enzyme activity. Detection kits (Solarbio, Beijing) were used to measure the activities of soil sucrase (S-SC), soil alkaline phosphatase (S-AKP/ALP), soil urease (S-UE) and soil polyphenol oxidase (S-PPO).
3. Crop growth indexes. After harvesting, crop samples were collected and divided into two parts, aboveground and underground, which were rinsed with distilled water and air-dried. The aboveground fresh weight (AFW) and underground fresh weight (UFW) were weighed using a balance. Subsequently, the samples were placed in a 105 °C oven for 15 min and were then dried to a constant weight at 75 °C, after which the samples were weighed to calculate the aboveground dry weight (ADW) and underground dry weight (UDW).
4. Physiological characteristics of crops. The chlorophyll a, chlorophyll b and total chlorophyll contents were measured using a chlorophyll content detection kit (Solarbio, Beijing, China). The content of soluble protein was determined by the Coomassie brilliant blue G-250 method, and the catalase (CAT) activity, superoxide dismutase (SOD) activity, peroxidase (POD) activity and malondialdehyde (MDA) content in the plants was determined by UV absorption, nitrogen blue tetrazolium (NBT) photoreduction, guaiacol colourimetry, and thiobarbituric acid methods.
5. Integrated biological response version 2 (IBRv2), proposed by Sanchez *et al.* (2013), was applied to explain the effect of irrigation management on arid soil enzyme activities (Diaz *et al.* 2021). The IBRv2 index was calculated using the main indexes, i.e., soil and crop, following Equation (1). Higher IBRv2 scores indicated larger deviations for all selected factors with regard to the chosen reference level.

$$\mathrm{IBRv2} = \sum_{i=1}^{n} |A_i| \tag{1}$$

where A_i, calculated using Equation (2), is a deviation value for each factor with respect to that measured in the reference plots.

$$A_i = \left[\frac{Log\dfrac{x_i}{x_0} - \mu}{\sigma}\right] - Z_0 \tag{2}$$

For each factor, single data points (x_i) were divided by the corresponding reference value (x_0) and were log-transformed to reduce the variance. The log-transformed data were standardized considering the general mean (μ) and the standard deviation (σ). Then, A_i, obtained by subtracting the mean of the reference standardized data (Z_0) from the mean of the standardized value, was plotted in a radar graph to visually check for the induction (positive A_i or area in the radar plot above the reference level) or inhibition (vice versa) of each factor.

2.4. Data analysis

Excel 2010 software was used to calculate the experimental data, SPSS 25.0 software was used to conduct univariate analysis of variance, followed by the least significant difference (LSD) test ($P < 0.05$).

3. RESULTS

3.1. Effects of mixed irrigation on soil

3.1.1. Effects on soil water and salt

Soil moisture is an important factor in soil fertility and is the main source of crop water absorption. The soil salt content is the main parameter of soil salinity and the main index of soil salinization. Soil salinity is highly positively correlated with $EC_{1:5}$, which is easy to determine. Generally, the $EC_{1:5}$ value represents the soil salt content. The variations in the soil water content and $EC_{1:5}$ under mixed irrigation with brackish water and reclaimed water after crop harvesting are shown in Figure 1.

As shown in Figure 1, compared with CK, the soil moisture content increased slightly by 10.30% in T7, but this was not significant. At the same salinity of brackish water, the soil moisture content decreased gradually with the increase in the proportion of reclaimed water in the mixture. For example, when the salinity of brackish water was 3 g/L, there was no significant difference between T1 and T3 or T3 and T5 ($P > 0.05$), but there was a significant difference between T1 and T5 ($P < 0.05$); when the salinity of brackish water was 5 g/L, T2 was significantly different from T4 and T6 ($P < 0.05$), but there was no significant difference between T4 and T6 ($P > 0.05$). At the same mixing ratio of brackish water and reclaimed water, the higher the salinity, the greater the soil moisture content, and there was no significant difference among the other mixed treatments ($P > 0.05$) except for the significant difference between the two brackish water irrigation treatments. Therefore, there were no obvious impacts of reclaimed water and freshwater irrigation on the soil water content, and the soil moisture content decreased gradually with the increase in the proportion of reclaimed water in the mixture.

When the salinity of brackish water was constant, the $EC_{1:5}$ in different mixed irrigation treatments was significantly higher than that of CK ($P < 0.05$), and soil salinity increased obviously with the increase in the proportion of reclaimed water in the mixture ($P < 0.05$). At the same mixing ratio of brackish water and reclaimed water, there was a positive correlation between $EC_{1:5}$ and the salinity of brackish water, and the difference between treatments was significant ($P < 0.05$). Therefore, soil salinity was mainly determined by the salt content in irrigation water. This was consistent with the changing trend of the soil moisture content because the higher the salt content, the stronger the limiting effect on crop water absorption, resulting in more water remaining in the soil.

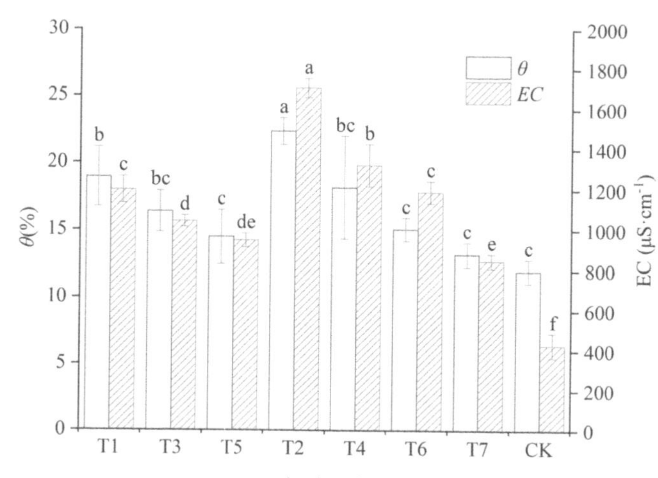

Figure 1 | Variations of soil water and salt content after mixed irrigation with brackish water and reclaimed water. *Note:* θ represents soil water content; different lower case letters on boxplots represent the significant differences at the level of 0.05. The same as below.

3.1.2. Effects on soil water-soluble ions

Soil water-soluble salt is an important attribute of saline-alkali soil and a restraint on crop growth. In addition to the determination of the pH value and total salt content, the analysis of soil water-soluble salt also includes the determination of anions (CO_3^{2-}, HCO_3^-, Cl^-, SO_4^{2-}) and cations (K^+, Na^+, Ca^{2+}, Mg^{2+}). The changes in the water-soluble ions in the soil after mixed irrigation with brackish water and reclaimed water are shown in Figure 2.

Figure 2 shows (1) the same change trends for the soil K^+, Ca^{2+} and SO_4^{2-} contents, and the contents of soil K^+, Ca^{2+} and SO_4^{2-} in T7 were significantly higher ($P < 0.05$) than those in CK (except for the Ca^{2+} content). At the same salinity of brackish water, the contents of soil K^+, Ca^{2+} and SO_4^{2-} increased gradually with the increase in the proportion of reclaimed water in the mixture. For example, when the salinity of brackish water was 3 g/L, the contents of soil K^+, Ca^{2+} and SO_4^{2-} in T7 were the highest. The soil K^+ content in T7 was significantly different from those of T1 and T5 ($P < 0.05$) but not significantly different from that of T3, while the contents of Ca^{2+} and SO_4^{2-} were significantly higher in T7 than in the other treatments ($P < 0.05$). When the mixing ratio of brackish water and reclaimed water was constant, the contents of K^+, Ca^{2+} and SO_4^{2-} in soil increased with the increase in the salinity of brackish water, but the difference was not significant ($P > 0.05$).

(2) The soil Na^+ and Cl^- contents exhibited the same trend, and the contents in T7 were significantly higher than those in CK ($P < 0.05$). At the same salinity of brackish water, the contents of soil Na^+ and Cl^- decreased with the increase in the proportion of reclaimed water in the mixture, and the difference between different treatments was significant ($P < 0.05$) (but there was no significant difference between the T3 and T5 treatments). At the same ratio of brackish water to reclaimed water, the contents of soil Na^+ and Cl^- increased significantly with increasing salinity of brackish water ($P < 0.05$).

(3) With respect to the contents of Mg^{2+} and HCO_3^- in soil, CK had higher values than T7, and the difference in the soil Mg^{2+} content reached a significant level ($P < 0.05$). Under the same salinity of brackish water, the soil Mg^{2+} content

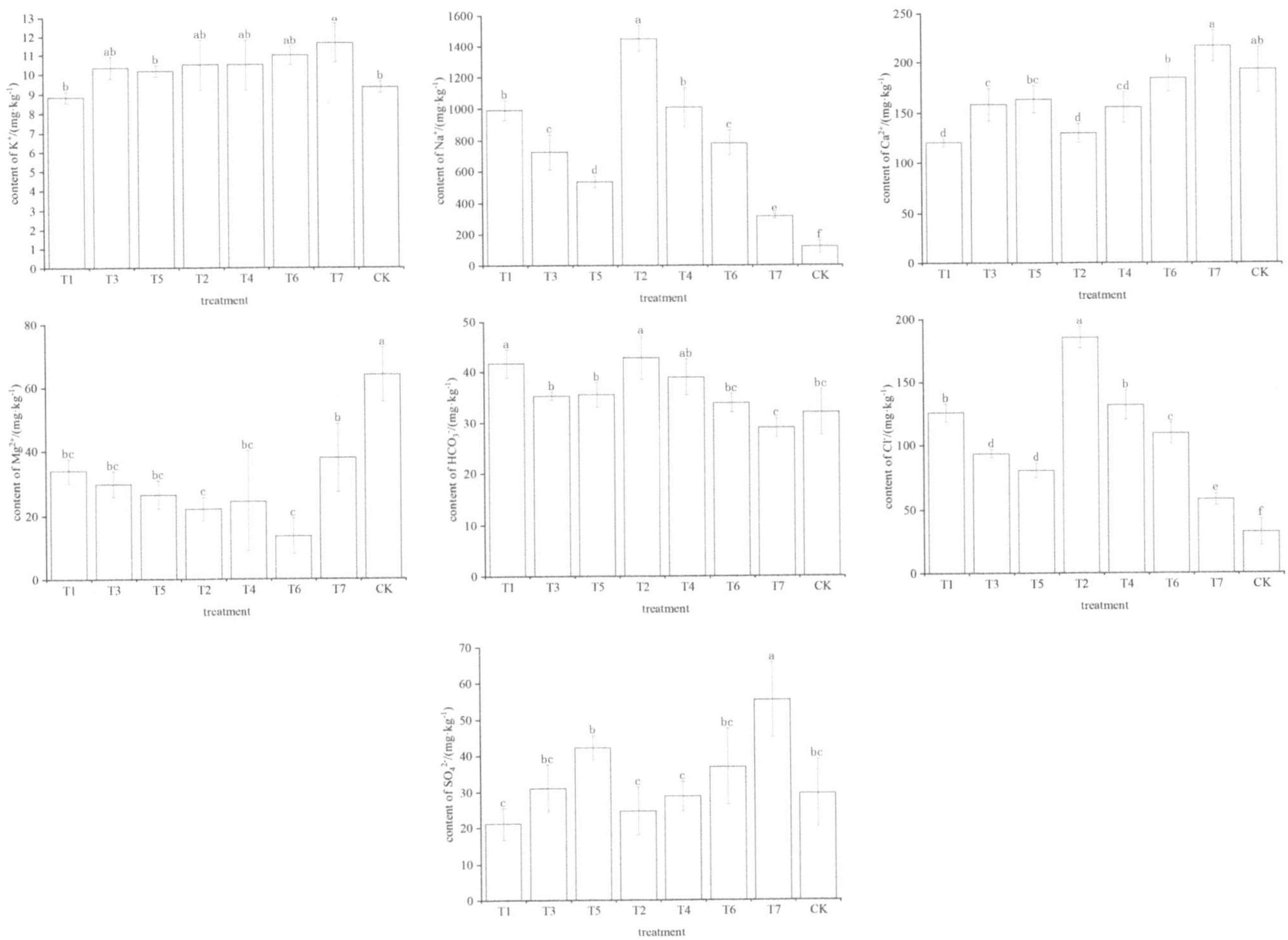

Figure 2 | Variations of soil water-soluble ions content after mixed irrigation with brackish water and reclaimed water.

decreased gradually with the increase in the proportion of reclaimed water in the mixture, but the difference was not significant ($P > 0.05$) (except in T7). The soil HCO_3^- content also decreased gradually, and the soil HCO_3^- content in brackish water was higher than that in reclaimed water, but there was no significant difference between the 1:1 and 1:2 mixtures of brackish water and reclaimed water ($P > 0.05$).

3.1.3. Effects on soil enzyme activity

Soil phosphatase is an enzyme that catalyses the mineralization of soil organophosphorus compounds. Its activity level directly affects the decomposition and transformation of organophosphorus in soil and its bioavailability. In northern China, soil is alkaline, so soil phosphatase activity in the test refers to alkaline phosphatase (S-AKP/ALP). S-PPO is mainly derived from the decomposition and release of soil microorganisms, plant root exudates and animal-plant residues. This enzyme catalyses the oxidation of aromatic compounds in soil into quinone, which reacts with proteins, amino acids, carbohydrates, minerals and other substances in soil to generate organic matter and pigments and completes the cycling of soil aromatic compounds for soil environmental remediation. The enzymatic products of S-SC are closely related to the contents of organic matter, nitrogen and phosphorus in the soil, the abundance of microorganisms and the soil respiration intensity. S-UE activity is positively correlated with the soil microbial abundance, organic matter content, total nitrogen and available nitrogen, reflecting the nitrogen status of soil. The changes in the activities of S-AKP/ALP, S-PPO, S-SC and S-UE after mixed irrigation with brackish water and reclaimed water are shown in Figure 3.

As shown in Figure 3, there was no significant difference between T7 and CK in terms of S-AKP/ALP and S-PPO activity ($P > 0.05$). At the same salinity of brackish water, the activities of S-AKP/ALP and S-PPO changed, exhibiting a 'V'-shaped curve with the increase in the proportion of reclaimed water in the mixture, and the higher the salinity of brackish water, the earlier the inflection point. At the same ratio of brackish water to reclaimed water, there was no significant difference in

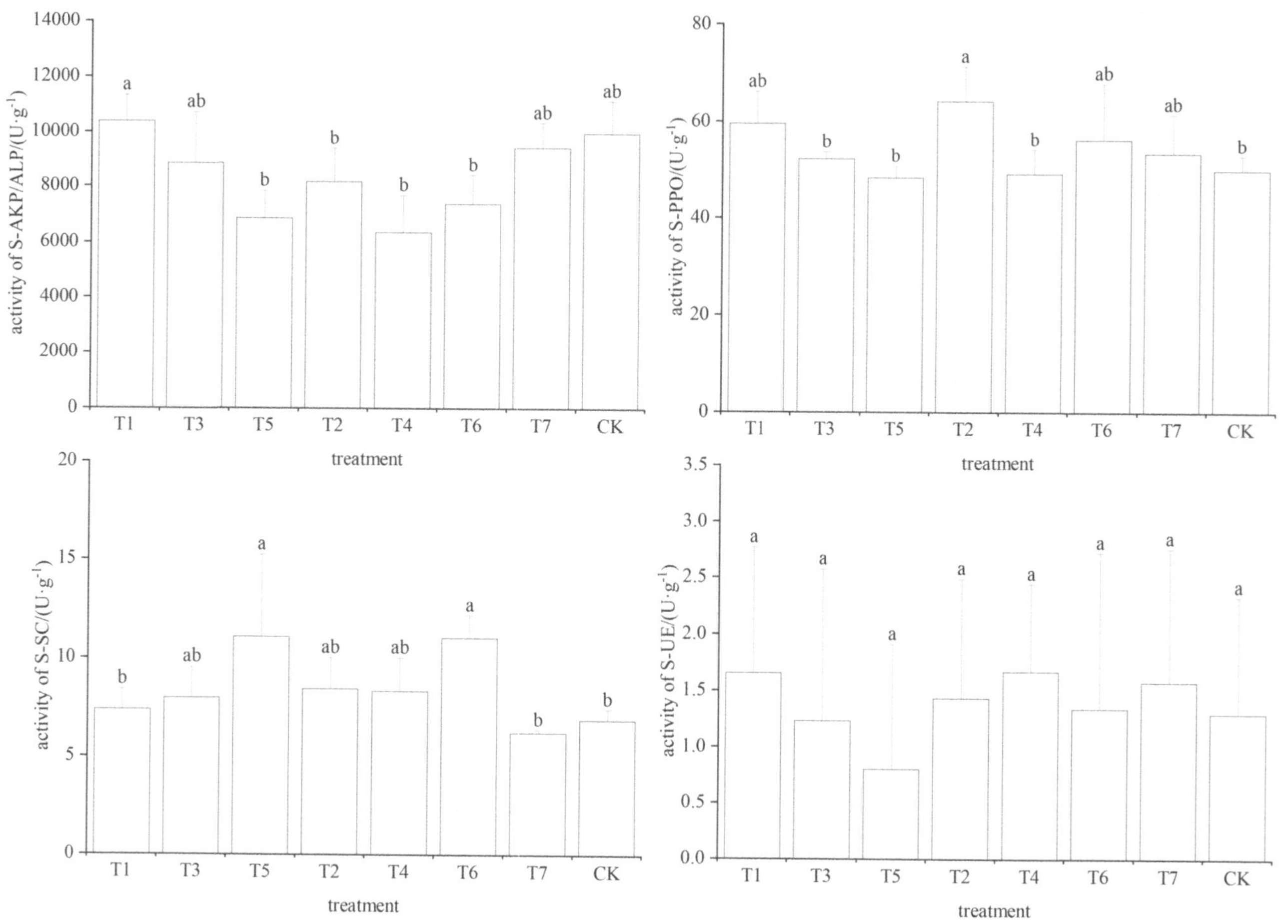

Figure 3 | Variations of soil enzyme activities after mixed irrigation with brackish water and reclaimed water.

S-AKP/ALP and S-PPO activities ($P > 0.05$) except that the activity of S-AKP/ALP decreased significantly ($P < 0.05$) with the increase in salinity under brackish water irrigation.

There was no significant difference in S-SC activity between T7 and CK ($P > 0.05$). At a given salinity of brackish water, the S-SC activity showed an initial increase followed by a decrease with the increase in the proportion of reclaimed water in the mixture, and the activity was the highest at a brackish water/reclaimed water ratio of 1:2, which was significantly higher ($P < 0.05$) than that of reclaimed water irrigation and brackish water irrigation (but brackish water with a salinity of 3 g/L was not significantly different from reclaimed water irrigation and brackish water irrigation). At the same mixing ratio of brackish water and reclaimed water, the S-SC activity increased with the increase in the proportion of reclaimed water in the mixture, but the difference was not significant ($P > 0.05$). In addition, there was no significant difference in S-UE activity among the different treatments ($P > 0.05$) and there was no obvious change trend.

3.1.4. Effects on SOM

SOM is one of the important indexes of soil fertility, in addition to basic conditions to ensure the normal growth of plants, and an important measure to increase production. The activities of microorganisms, the synthesis and decomposition of organic matter, the ability of soil to retain nutrients and the availability of elements in soil are all related to soil pH. The variations in the SOM content and pH after mixed irrigation with brackish water and reclaimed water are shown in Figure 4.

As seen from Figure 4, no significant difference in soil pH was found between T7 and CK, and soil pH decreased slightly with the increase in the proportion of reclaimed water in the mixture at the same salinity of brackish water. The soil pH in T4 was significantly higher than that in T6 and T7 ($P < 0.05$) when the salinity of brackish water was 5 g/L, while the soil pH decreased (but not significantly) with the increase in the salinity of brackish water at the same mixing ratio of brackish water and reclaimed water.

The SOM in T7 was slightly higher than that of CK, but the difference was not significant ($P > 0.05$). At the same mixing ratio of brackish water and reclaimed water, SOM decreased with increasing salinity of brackish water, and the difference reached a significant level ($P < 0.05$) when the mixing ratio of brackish water and reclaimed water was 1:2. With the increase in the proportion of reclaimed water in the mixture, SOM increased first and then decreased in response to brackish water with a salinity of 3 g/L, while SOM decreased gradually in response to brackish water with a salinity of 5 g/L. Therefore, regardless of the salinity of brackish water, the SOM in T7 was significantly lower than that in the other treatments ($P < 0.05$).

3.1.5. Effects on soil WDPT

Soil water repellency refers to the physical phenomenon in which water cannot or has difficulty moistening the surface of soil particles (Yang *et al.* 1994). Soil water repellency will lead to an uneven distribution of soil water, soil water loss and dry soil surfaces, and enhanced surface runoff and soil erosion after rainfall or irrigation and will thus hinder the growth of crops (Dekker & Jungerius 1990). The strength of soil water repellency is generally characterized by WDPT. A WDPT that exceeds 5 seconds indicates soil water repellency. The changes in soil WDPT after mixed irrigation with brackish water and reclaimed

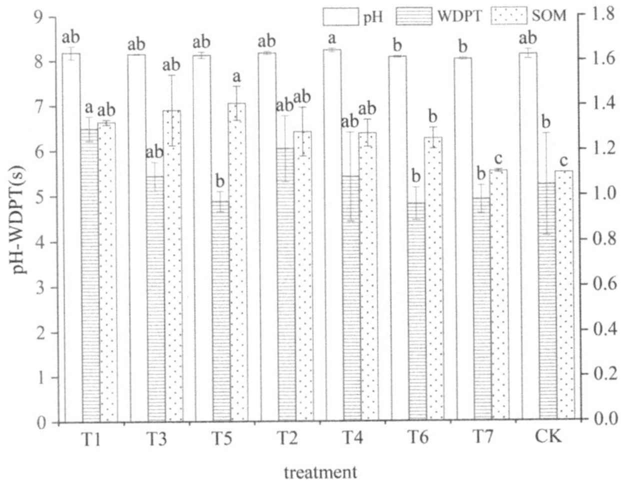

Figure 4 | Variation of soil organic matter, pH value and WDPT under mixed irrigation with brackish water and reclaimed water.

water are shown in Figure 4. The WDPT in T7 was slightly lower than that in CK, but the difference was not significant ($P > 0.05$). At the same salinity of brackish water, WDPT decreased with the increase in the proportion of reclaimed water in the mixture, and WDPT in T1 was significantly higher ($P < 0.05$) than that in T5 and T7 in response to brackish water with a salinity of 3 g/L. The higher the salinity of brackish water, the higher the WDPT, but there was no significant difference at the same ratio of brackish water to reclaimed water ($P > 0.05$).

3.2. Effects of mixed irrigation on crops

3.2.1. Crop growth

Biomass plays a very important role in the formation of ecosystem structure and function. The changes in aboveground biomass and underground biomass (fresh weight and dry weight) of crops after mixed irrigation with brackish water and reclaimed water are shown in Figure 5.

As shown in Figure 5, there were no significant differences in AFW and ADW between T7 and CK ($P > 0.05$). At the same salinity of brackish water, AFW and ADW under brackish water irrigation were significantly lower than those under reclaimed water irrigation ($P < 0.05$), while AFW and ADW under mixed irrigation were generally higher than those under brackish water irrigation, but the difference was not significant ($P > 0.05$). At a given mixing ratio of brackish water to reclaimed water, the higher the salinity of brackish water, the lower the AFW and ADW; significant differences in AFW were observed ($P < 0.05$) between brackish water irrigation and 1:1 mixed irrigation with brackish water-reclaimed water, but there were no other significant treatment differences ($P > 0.05$). Regardless of the different salinities of brackish water or the ratio of brackish water to reclaimed water, no significant treatment differences were observed for UFW and UDW ($P > 0.05$).

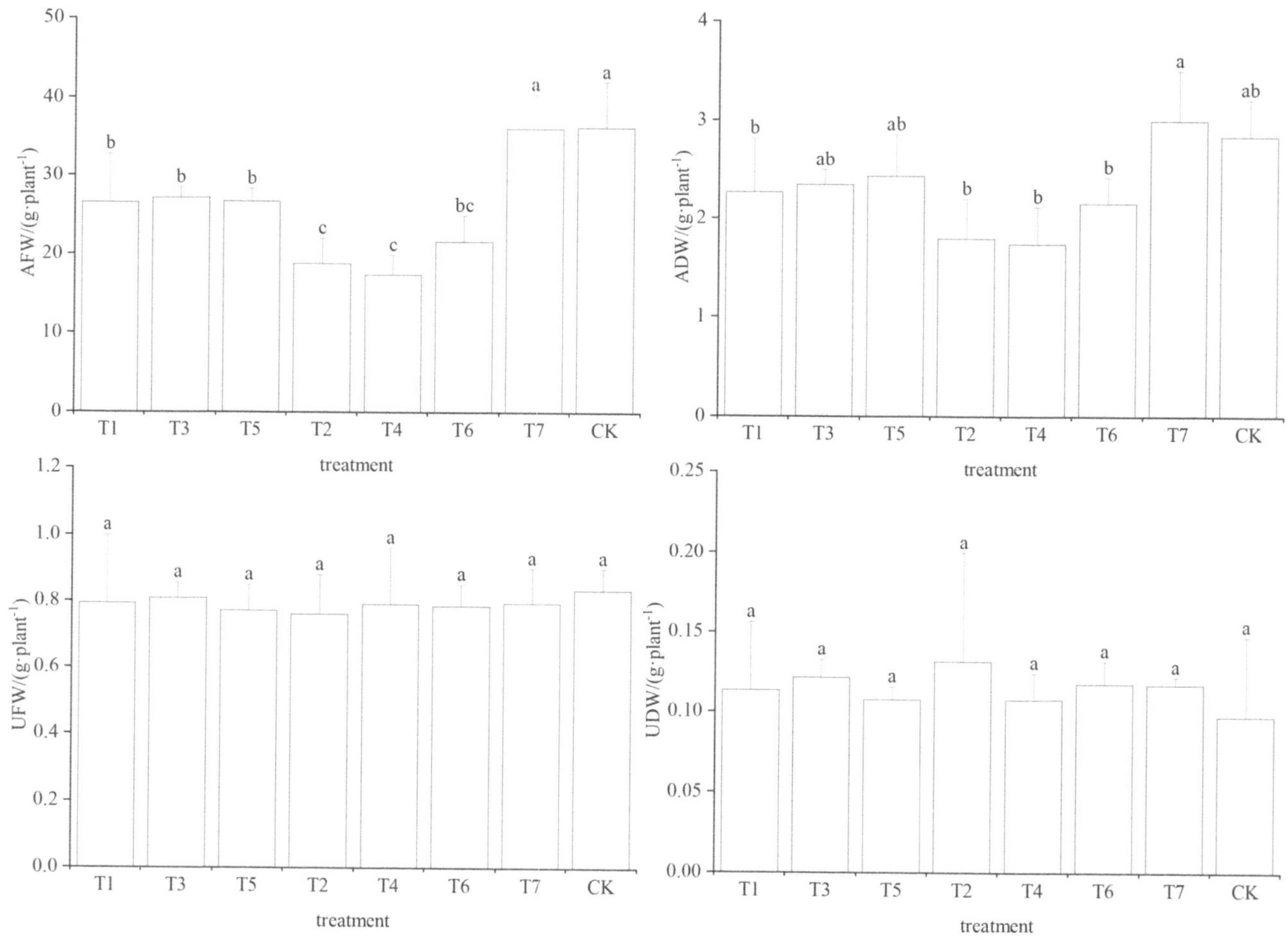

Figure 5 | Variations of biomass of crop after mixed irrigation with brackish water and reclaimed water.

3.2.2. Physiological and biochemical characteristics of crops

(1) Chlorophyll content. The chlorophyll content, which is closely related to photosynthesis and nutritional status, is an important indicator of plant growth. The chlorophyll in the chloroplasts of higher plants mainly includes chlorophyll a and chlorophyll b. The contents of chlorophyll a, chlorophyll b and total chlorophyll of leaves after mixed irrigation with brackish water and reclaimed water are shown in Figure 6.

As observed in Figure 6, there were no significant differences in chlorophyll a, chlorophyll b or total chlorophyll among all treatments ($P > 0.05$), and their change trends were consistent. T7 had chlorophyll a, chlorophyll b and total chlorophyll contents that were 3.97%, 5.61% and 4.47% higher, respectively, than those of CK. Compared with T7, under 1:2 and 1:1 mixed irrigation and brackish water irrigation, the chlorophyll a content decreased by 4.98%–7.63%, 3.82%–7.85% and 5.15%–9.26%, the content of chlorophyll b decreased by 10.88%–13.99%, 8.2%–12.73% and 9.28%–13.46%, and the total chlorophyll content decreased by 9.61%–9.76%, 6.12%–9.37% and 6.43%–10.15%, respectively. Therefore, mixed irrigation with brackish water and reclaimed water had no significant effect on the chlorophyll content of the crops.

(2) Enzyme activity in the leaf. Antioxidant enzymes, such as SOD, POD and CAT, can convert peroxides into less toxic or harmless substances. SOD, which scavenges superoxide anion free radicals by disproportionation reactions, plays an important role in biological antioxidant systems. POD can eliminate the toxicity of hydrogen peroxide, phenols and amines. CAT plays an important role in the active oxygen scavenging system. The change in the antioxidant enzyme activity in the crop under mixed irrigation of brackish water and reclaimed water is shown in Figure 7.

As shown in Figure 7, no significant difference occurred in SOD and CAT activities between T7 and CK ($P > 0.05$). At the same salinity of brackish water, the activities of SOD and CAT first increased and then decreased with the increase in the proportion of reclaimed water in the mixture, reaching a maximum at a 1:2 ratio of brackish water to reclaimed water. At 3 g/L salinity of brackish water, SOD activity in T7 was significantly lower than that in T3 and T5 ($P < 0.05$), but there was no significant difference between T1, T3 and T5 ($P > 0.05$); there was no significant difference in CAT activity between treatments ($P > 0.05$). When the salinity of brackish water was 5 g/L, SOD activity in T6 was significantly higher than that in other treatments ($P < 0.05$), CAT activity in T6 was significantly higher than T2 ($P < 0.05$), and there were no significant differences in SOD and CAT activity between T2 and T4 ($P > 0.05$), but SOD activity in T6 was significantly higher than that in T7 ($P < 0.05$). At a given mixing ratio of brackish water to reclaimed water, the activities of SOD and CAT increased with increasing salinity of brackish water, but the difference did not reach a significant level ($P > 0.05$).

There was no significant difference in POD activity among the different treatments ($P > 0.05$), but the change trend was different under different salinities of brackish water. For example, mixed irrigation improved POD activity to a certain extent compared to brackish water irrigation at 3 g/L salinity, while the opposite trend occurred at 5 g/L salinity.

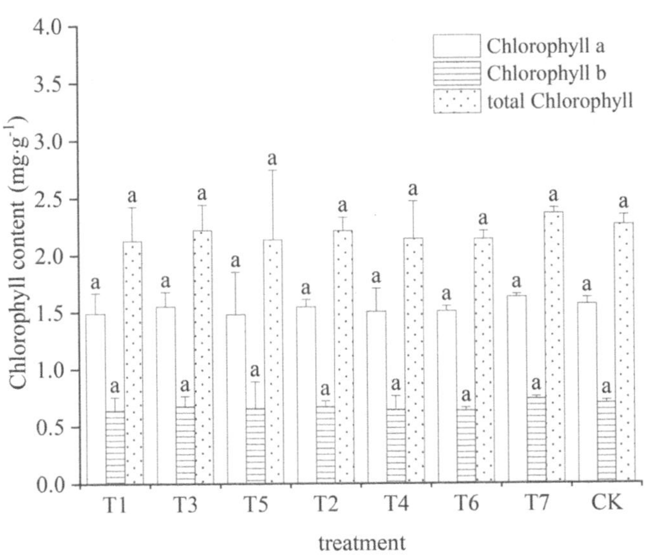

Figure 6 | Variations of chlorophyll contents of crop under mixed irrigation with brackish water and reclaimed water.

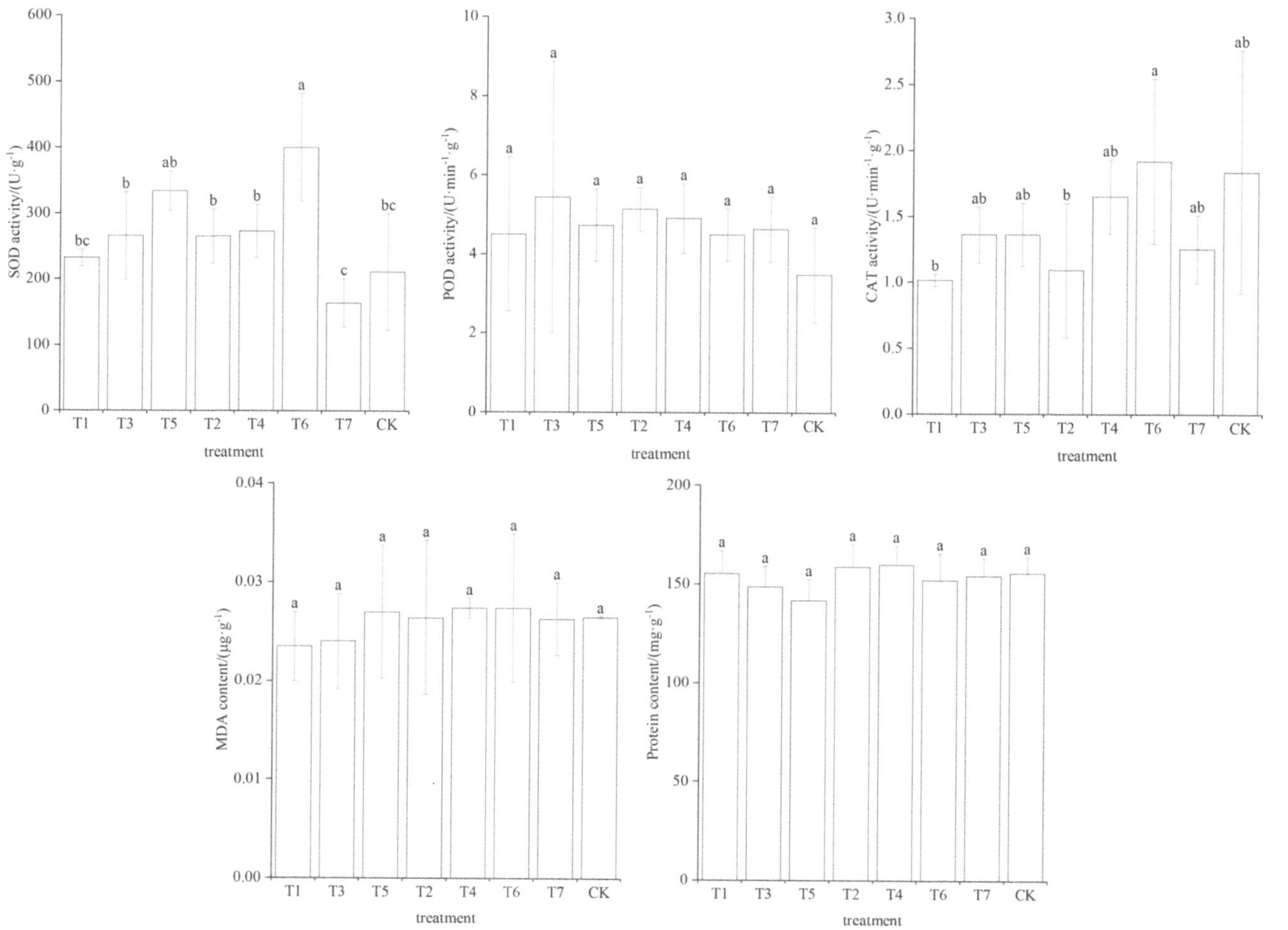

Figure 7 | Variation of leaf enzyme activities after mixed irrigation with brackish water and reclaimed water.

(3) MDA. The production of MDA aggravates membrane damage. Understanding the degree of membrane lipid peroxidation by MDA can indirectly determine the degree of damage caused to the membrane system and the stress resistance of plants. The change in the MDA content of crops after mixed irrigation with brackish water and reclaimed water is shown in Figure 8. There was no significant difference in the MDA content among the different treatments ($P > 0.05$). At the same salinity of brackish water, the MDA content increased with the increase in the proportion of reclaimed water in the mixture but decreased slightly under reclaimed water irrigation; at the same ratio of brackish water to reclaimed water, the higher the salinity of brackish water, the higher the MDA content.

(4) Soluble protein. Soluble proteins are important osmotic regulators and nutrients. Their increase and accumulation can improve the water retention capacity of cells and protect the life substances and biofilms of cells. The change in the soluble protein content of crops under mixed irrigation of brackish water and reclaimed water is shown in Figure 8. At the same salinity of brackish water, the soluble protein content decreased with the increase in the proportion of reclaimed water in the mixture, and the content increased slightly under reclaimed water irrigation. At the same mixing ratio of brackish water to reclaimed water, the content of soluble protein was positively correlated with the salinity of brackish water. However, there was no significant difference in the soluble protein content among the different treatments ($P > 0.05$).

3.3. Evaluation of mixed irrigation in a soil-crop system based on IBRv2

IBRv2 was first used to evaluate the toxicity of environmental pollution to organisms and then gradually applied to explain the effect of irrigation on enzyme activity. The effect of mixed irrigation with brackish water and reclaimed water on the soil-crop system was evaluated by IBRv2. Based on the above results, ten crop and soil indexes (θ, $EC_{1:5}$, AFW, SOM, WDPT, S-AKP/ALP, S-PPO, S-SC, CAT, SOD), with a relatively obvious influence and no repeated affecting factors, were selected

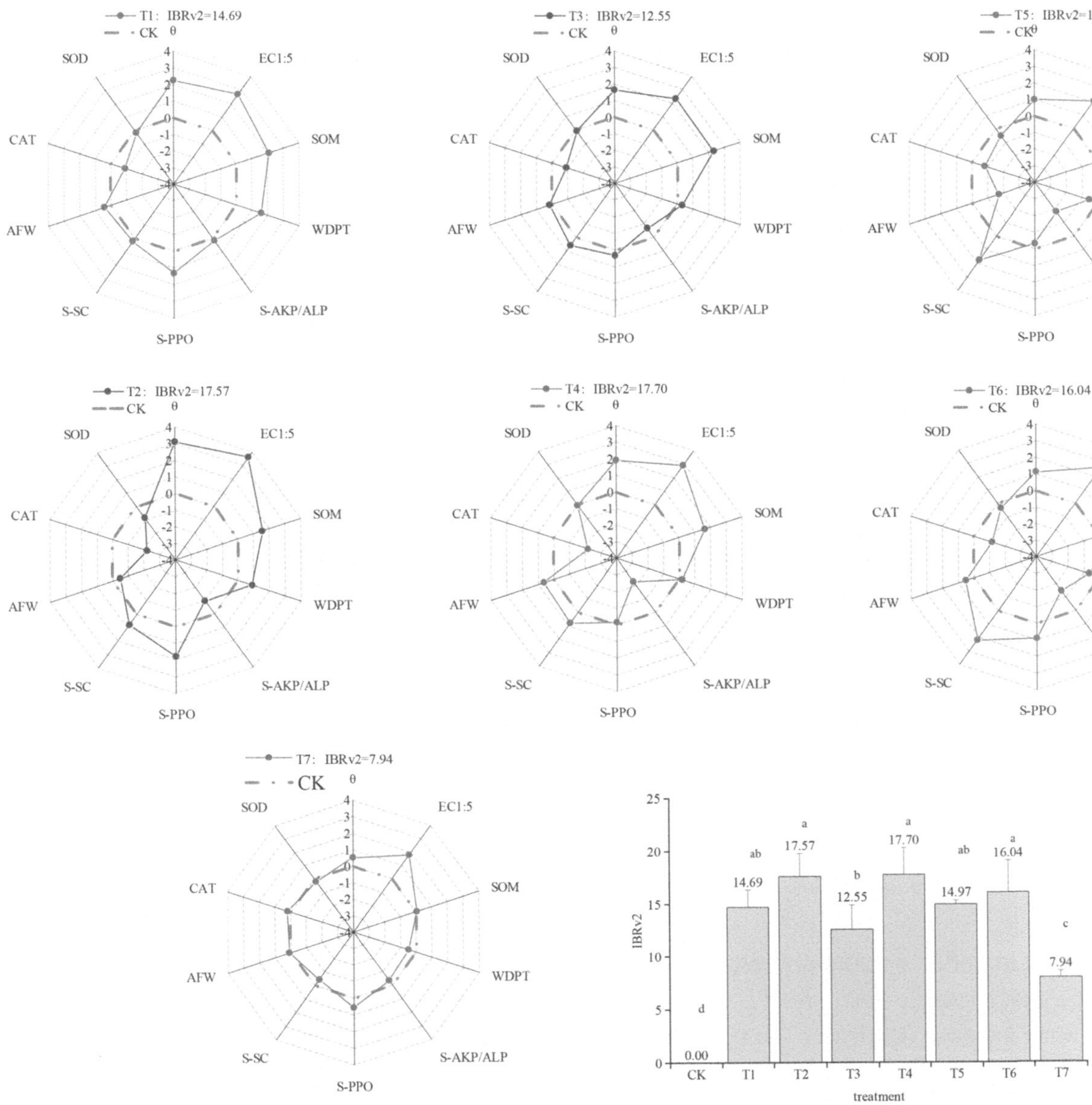

Figure 8 | IBRv2 values and associated radar plots under different irrigation treatments.

to calculate the IBRv2 value with the measured value of CK as the control reference. The changes in IBRv2 and the radar map are shown in Figure 8.

As seen in Figure 8, all treatments had activation effects on θ, $EC_{1:5}$, SOM, S-PPO and S-SC and certain inhibitory effects on CAT, SOD, S-AKP/ALP and AFW as a whole except for certain inhibitory effects on S-SC under T7, S-PPO under T4 and T5 and activation effects on CAT under T7. WDPT was activated under brackish water irrigation and 1:1 mixed irrigation with brackish water and reclaimed water but was inhibited under reclaimed water irrigation and 1:2 mixed irrigation of brackish water and reclaimed water.

The IBRv2 value in T7 was the smallest and significantly lower than other treatments ($P < 0.05$), indicating that the deviation between reclaimed water irrigation and freshwater irrigation was the smallest. However, the combined utilization of brackish water and reclaimed water could be considered when using brackish water for irrigation in areas where freshwater resources are insufficient due to the constraints of reclaimed water resources such as limited daily discharge and imperfect reclaimed water networks. At brackish water salinities of 3 g/L and 5 g/L, the most appropriate IBRv2 occurred in T3 and T6, with scores of 12.55 and 16.04, respectively. Therefore, the 1:1 ratio of mixed irrigation of brackish water and reclaimed water was suitable at 3 g/L of the salinity of brackish water but the 1:2 ratio at 5 g/L of the salinity of brackish water.

4. DISCUSSION

4.1. Effects of mixed irrigation on soil water, salt and water-soluble ions

According to previous results, the soil moisture content and electrical conductivity increased with increasing salinity of brackish water at the same soil depth (Yang *et al.* 2020b). In addition, the soil salt content increased with increasing salinity of brackish water (Zhang *et al.* 2020a). Soil moisture increased with increasing salinity of brackish water (1–6 g/L) but decreased above a certain salinity (9–12 g/L) (Yang *et al.* 2020a). The results in the current paper show that the soil moisture content and soil salt content decreased gradually with the increase in the proportion of reclaimed water in the mixture, which is consistent with previous research results. The reason is that the salinity of reclaimed water was higher than that of freshwater, the salinity in the mixture decreased with the increase in the proportion of reclaimed water in the mixture, and the soil salt content was lower after irrigation. Soil salinity will inhibit crop water absorption to a certain extent, so at the same irrigation amount, the soil moisture content will be higher due to the decrease in crop water absorption with the increase in salinity.

The higher the soil electrical conductivity, the higher the concentration of soluble ions in the soil and the higher the total salt content. Soil electrical conductivity can represent soil salinity, and different soil salinities will affect the content of salt ions (Zamanian *et al.* 2016). In this paper, the results showed that after crop harvest, the soil K^+, Ca^{2+} and SO_4^{2-} contents generally exhibited the same trend, and their contents under reclaimed water irrigation were all significantly higher than those under freshwater irrigation (except for the Ca^{2+} content). The ion content tended to increase gradually with the increase in the proportion of reclaimed water in the mixture at the same salinity of brackish water, while the contents of K^+, Ca^{2+} and SO_4^{2-} in soil exhibited an increasing trend with the increase in the salinity of brackish water at the same mixing ratio of brackish water to reclaimed water. The reason is the relatively high contents of K^+, Ca^{2+} and SO_4^{2-} in reclaimed water. The results also showed that the contents of soil Na^+ and Cl^- increased significantly under mixed irrigation, and the contents of soil Na^+ and Cl^- decreased significantly with the increase in the proportion of reclaimed water in the mixture, which may be because the changes in soil Na^+ and Cl^- were mainly related to the contents of Na^+ and Cl^- in irrigation water and water absorption by roots. Na^+ and Cl^- move with water to the root, and water is absorbed by the root. However, less Na^+ and Cl^- were absorbed by the roots and then accumulated in the soil around the roots. In addition, the contents of soil Ca^{2+} and Mg^{2+} under mixed irrigation were lower than those under freshwater irrigation and reclaimed water irrigation, indicating that mixed irrigation may promote the absorption of Ca^{2+} and Mg^{2+} by crops, which may be related to the competitive adsorption of ions, but the reasons need to be further discussed.

4.2. Effects of mixed irrigation on SOM and WDPT

Soil organic matter is an important index of soil fertility. The content of SOM was reported to increase, but not significantly, under reclaimed water irrigation (Chen *et al.* 2014), which is consistent with the results in this paper. However, it was found that the SOM content under reclaimed water irrigation was significantly higher than that under freshwater irrigation (Han *et al.* 2018b), which is slightly different from the results of this experiment. In this paper, the SOM content increased slightly with no significant difference under reclaimed water irrigation. The reason may be due to the differences in the quality of reclaimed water or that crops were planted in our experiment, while bare land irrigation was used in the previous experiment (Han *et al.* 2018b). Our results also showed that the SOM content under mixed irrigation was higher than that under reclaimed water irrigation, which may be due to the high salt content. A high salt content would reduce organic matter and the absorption of nutrients by crops; thus, the SOM was relatively higher after crop harvest.

Soil water repellency is ubiquitous. Extracellular polymers are formed by soil microorganisms under reclaimed water, and soil water repellency may be caused by the cover of extracellular polymers on the surface of soil mineral particles or soil aggregates (Morales *et al.* 2010). It was reported that the WDPT in the profile of four different soil types increased under reclaimed water irrigation (Shang *et al.* 2012), which slightly differed from the experimental results in this paper; that is, soil WDPT under reclaimed water irrigation was slightly lower than that under freshwater irrigation, and the reason may be due to the different components in the reclaimed water or soil texture types. Shang *et al.* (2012) compared the WDPT of soil after infiltration with that of air-dried soil before experiments based on indoor soil column tests. In the current experiment, however, changes in soil WDPT under reclaimed water and freshwater irrigation were compared, and different soil moisture contents and irrigation processes may also have effects on soil WDPT. Liu *et al.* (2011) found that weak water repellency occurred in many loamy soil profiles after infiltration with 3.5 g/L brackish water in Yangling in Shaanxi and in saline-alkali soil in Manas in Xinjiang, especially the latter. This is consistent with the experimental results that the soil WDPT increased with the increase in the proportion of brackish water in the mixture. A positive correlation was found between WDPT and the SOM content (Ren *et al.* 2017), but there was no perfect positive correlation in our results, which was closely related to the complex water quality. Our results also showed that compared with brackish water irrigation, WDPT decreased with the increase in the proportion of reclaimed water in the mixture. Related studies on mixed irrigation with brackish water and reclaimed water have not been reported, and this paper only preliminarily discussed mixed irrigation. Therefore, the related theoretical mechanism of mixed irrigation with brackish water and reclaimed water needs to be discussed further.

4.3. Effects of mixed irrigation on soil enzyme activity

There were some differences in the effects of reclaimed water irrigation on soil enzyme activities. Reclaimed water irrigation had no significant effect on soil enzyme activity (Ndour *et al.* 2008). Soil urease, alkaline phosphatase and sucrase activities increased to some extent under reclaimed water irrigation, but the difference was not significant compared with the control (Pan *et al.* 2012). Soil enzyme activity was affected by soil nutrients and heavy metal pollution (Zhang *et al.* 2006), and the timing of reclaimed water irrigation had different impacts on soil enzyme activities (Guo *et al.* 2012). In this paper, the effects of reclaimed water irrigation on soil enzyme activities (alkaline phosphatase, sucrase, urease and polyphenol oxidase) were not significant, which is basically consistent with previous research results. In addition, mixed irrigation with different ratios of brackish water to reclaimed water had a certain effect on soil enzyme activity, i.e., mixed irrigation with a 1:2 ratio of brackish water and reclaimed water enhanced soil sucrase activity, and mixed irrigation with a 1:1 ratio of 5 g/L brackish water and reclaimed water enhanced urease activity to some extent. Therefore, the best appropriate mixing ratio of brackish water to reclaimed water can improve soil enzyme activity, but the specific ratio and the regulatory mechanism of mixed irrigation with brackish water and reclaimed water on soil enzyme activity need to be further discussed.

4.4. Effects of mixed irrigation on the biomass and physiological characteristics of crops

Biomass is one of the important indicators of crop yield, especially that of vegetables. The results in this paper showed that the mixed irrigation of brackish water-reclaimed water had no significant effect on the aboveground dry weight, but the aboveground fresh weight was significantly lower compared with reclaimed water irrigation, which was consistent with the change in the soil moisture. The soil moisture content under mixed irrigation was higher than that under freshwater irrigation, indicating that the lower the amount of water absorbed by crops under the same irrigation amount and meteorological conditions, the lower the plant water content. Therefore, the fresh weight decreased significantly, but there was no significant difference in the dry weight. In previous research, compared with freshwater irrigation, reclaimed water irrigation significantly improved the yield of fruits and vegetables (Wu *et al.* 2010). The photosynthesis and yield indexes of cucumber irrigated with reclaimed water were higher than those irrigated with tap water (Wang *et al.* 2020). Our results also showed that there was no significant difference in biomass between reclaimed water irrigation and freshwater irrigation, which may be related to the quality and composition of the reclaimed water in different areas. The fresh fruit yield of processing tomato generally exhibited low yield under high salt levels (Zhang *et al.* 2020b), which is basically consistent with the results in this paper. It was found that the aboveground biomass (fresh weight and dry weight) of sweet and crisp pea decreased significantly with the increase in the salinity of irrigation water ($P < 0.05$) (Feng *et al.* 2020), which slightly differed from the result in this paper due to the difference in crop type or the difference between substrate cultivation and soil cultivation.

The chlorophyll content of leaves is an important index to represent plant growth. At the same irrigation quota, the SPAD value under reclaimed water irrigation was 5.54% higher than that under tap water irrigation (Wang *et al.* 2020). Reclaimed water irrigation could significantly increase the chlorophyll content of turfgrass (Lei & Li 2020), which is similar to the experimental results of this paper, that is, the content of chlorophyll under reclaimed water irrigation was higher than that under freshwater irrigation. Reclaimed water irrigation increased the MDA content and CAT activity of turfgrass (Lei & Li 2020), but our results showed no significant differences in the MDA content and CAT activity between reclaimed water and freshwater irrigation, which may be mainly due to the difference in water quality. The reclaimed water used in the former was a mixture of industrial and domestic effluent, while the effluent from domestic sewage treatment was used in this experiment. It was found that the POD and CAT activities and the MDA content of leaves increased significantly under recycled mine water irrigation (Yan *et al.* 2014), among which the change in the POD activity was similar to our experimental result, but the changes in the CAT activity and MDA content were slightly different. This may have been caused by different sources of reclaimed water or different water qualities or may be due to different crop types. No significant differences were found in the chlorophyll and MDA contents between brackish water irrigation and freshwater irrigation (Li *et al.* 2020), and

Li *et al.* (2016) also reported no significant differences in the chlorophyll and MDA contents between salt water irrigation (2–10 g/L) and freshwater irrigation, which is consistent with the experimental results in this paper.

5. CONCLUSIONS

The following conclusions were drawn:

1. At the same salinity of brackish water, the soil water content, salt content and WDPT decreased with the increase in the proportion of reclaimed water in the mixture of brackish water and reclaimed water. At the same ratio of brackish water to reclaimed water, the higher the salinity of brackish water, the higher the soil water content, salt content and WDPT.
2. Overall, soil cations changed from Na^{+} to Ca^{2+}, and anions changed from Cl^{-} to SO_4^{2-} with the increase in the proportion of reclaimed water in the mixture.
3. Different mixing ratios of brackish water to reclaimed water had different effects on soil enzyme activities. At the same salinity of brackish water, the activities of S-AKP/ALP and S-PPO changed, with an initial decrease followed by an increase with the increase in the proportion of reclaimed water in the mixture, and the higher the salinity of brackish water, the earlier the inflection point.
4. Mixed irrigation with brackish water and reclaimed water had a certain effect on the AFW of the crops, and the higher the salinity of brackish water, the more obvious the difference. Mixed irrigation had no significant effects on the ADW, underground biomass or crop physiological indexes (chlorophyll content, soluble protein content, MDA content, POD activity, CAT activity), but significantly improved SOD activity.
5. Based on the relative calculated results of IBRv2, the deviation of reclaimed water irrigation was the smallest, with a value of 7.94, followed by 1:1 (salinity of brackish water: 3 g/L) and 1:2 (salinity of brackish water: 5 g/L) mixed irrigation with brackish water and reclaimed water, with values of 12.55 and 16.04. Considering the soil environmental quality index, the limited daily reclaimed water amount and imperfect pipeline facilities, it is recommended that reclaimed water be considered an alternative source of freshwater to utilize with brackish water together in areas where freshwater resources are limited, and the suitable mixing ratio of brackish water to reclaimed water is 1:1 (at 3 g/L) or 1:2 (at 5 g/L).

ACKNOWLEDGEMENTS

This research was funded by the Scientific and Technological Project of Henan Province, grant number 202102110264, the Natural Science Foundation of Henan Province of China, grant number 202300410552 and Central Public-interest Scientific Institution Basal Research Fund, grant number FIRI202001-02, FIRI20210302, FIRI2019-04-02.

CONFLICT OF INTEREST

The authors declare no conflict of interest.

DATA AVAILABILITY STATEMENT

All relevant data are included in the paper or its Supplementary Information.

REFERENCES

Bhattacharyya, P., Tripathy, S., Chakrabarti, K., Chakraborty, A. & Banik, P. 2008 Fractionation and bioavailability of metals and their impacts on microbial properties in sewage irrigated soil. *Chemosphere* **72** (4), 543–550. doi:10.1016/j.chemosphere.2008.03.035.

Bustan, A., Cohen, S., De Malach, Y., Zimmermann, P., Golan, R., Sagi, M. & Pasternak, D. 2005 Effects of timing and duration of brackish irrigation water on fruit yield and quality of late summer melons. *Agricultural Water Management* **74** (2), 123–134. doi:10.1016/j.agwat.2004.11.009.

Chen, D., Wang, J., Guan, J., Xia, L. & Gong, X. 2014 Effects of reclaimed water irrigation on soil physicochemical properties and culturable microbial community. *Chinese Journal of Ecology* **33** (05), 1304–1311. doi:10.13292/j.1000-4890.20140326.001 (in Chinese with English abstract).

Cucci, G., Lacolla, G., Boari, F., Mastro, M. A. & Cantore, V. 2019 Effect of water salinity and irrigation regime on maize (*Zea mays* L.) cultivated on clay loam soil and irrigated by furrow in Southern Italy. *Agricultural Water Management* **222**, 118–124. doi:10.1016/j.agwat.2019.05.033.

Dekker, L. W. & Jungerius, P. D. 1990 Water repellency in the dunes with special reference to the Netherlands. *CATENA* **18**, 173–183.

Diaz, F. J., Sanchez-Hernandez, J. C. & Notario, J. S. 2021 Effects of irrigation management on arid soils enzyme activities. *Journal of Arid Environments* **185**, 104330. doi:10.1016/j.jaridenv.2020.104330.

Durán-Alvarez, J. C., Becerril-Bravo, E., SilvaCastro, V., Jiménez, B. & Gibson, R. 2009 The analysis of a group of acidic pharmaceuticals, carbamazepine, and potential endocrine disrupting compounds in wastewater irrigated soils by gas chromatography-mass spectrometry. *Talanta* **78** (3), 1159–1166. doi:10.1016/j.talanta.2009.01.035.

Feng, D., Zhang, J., Sun, C., Liu, X., Dang, H., Cao, C., Zheng, C., Sun, J., Li, K. & Ma, J. 2014 Effects of long-term irrigation with saline water on soil physical-chemical properties and activities of soil enzyme. *Journal of Soil and Water Conservation* **28** (3), 171–176. doi:10.13870/j.cnki.stbcxb.2014.03.032. (in Chinese with English abstract).

Feng, D., Zhu, Y. N., Zhou, T., Zhang, J. M. & Wang, Z. H. 2020 Effects of saline water irrigation on the growing and nutritional quality of sweet and crisp pea under substrate culture. *Journal of Irrigation and Drainage* **39** (2), 27–31. CSCD:6687050. (in Chinese with English abstract).

Gibson, R., Durán-Álvarez, J. C., León Estrada, K., Chávez, A. & Jiménez Cisneros, B. 2010 Accumulation and leaching potential of some pharmaceuticals and potential endocrine disruptors in soils irrigated with wastewater in the Tula Valley, Mexico. *Chemosphere* **81** (11), 1437–1445. doi:10.1016/j.chemosphere.2010.09.006.

Guo, X., Ma, T., Cui, Y., Chen, L., Du, P. & Liao, Y. 2012 Effect of sewage irrigation time on soil fertility and the enzyme activity. *Journal of Agro-Environment Science* **31** (4), 750–756. doi:CNKI:SUN:NHBH.0.2012-04-018. (in Chinese with English abstract).

Guo, W., Qi, X., Li, P., Li, Z., Zhou, Y. & Xiao, Y. 2017 Impact of reclaimed water irrigation and nitrogen fertilization on soil bacterial community structure. *Acta Scientiae Circumstantiae* **37** (1), 280–287. doi:10.13671/j.hjkxxb.2016.0189. (in Chinese with English abstract).

Han, Y., Li, P., Qi, X., Cui, B., Chang, D. & Ma, H. 2018a Effects of different levels of irrigation with reclaimed water on soil enzyme activity and distribution of thermotolerant coliforms. *Environmental Science* **39** (9), 4366–4374. doi:10.13227/j.hjkx.201801126. (in Chinese with English abstract).

Han, Y., Qi, X., Li, P., Chang, D., Li, Z., Cui, B. & Yan, X. 2018b Effects of the amount of irrigation with reclaimed wastewater on soil properties and distribution of pathogenic bacteria. *Journal of Irrigation and Drainage* **37** (8), 32–38. doi:10.13522/j.cnki.ggps.2017.0677. (in Chinese with English abstract).

Hassanli, A. M., Ebrahimizadeh, M. A. & Beecham, S. 2008 The effects of irrigation methods with effluent and irrigation scheduling on water use efficiency and corn yields in an arid region. *Agricultural Water Management* **96** (1), 93–99. doi:10.1016/j.agwat.2008.07.004.

Hu, Y. & Wu, W. 2018 Review and development strategy of irrigation with unconventional water resources in China. *Strategic Study of CAE* **20** (5), 69–76. (in Chinese with English abstract).

Hu, C., Li, P., Fan, L., Zhao, Z., Fan, T., Du, Z. & Qiao, D. 2013 Effects of reducing nitrogen topdressing under reclaimed water irrigation on quality and yield of greenhouse tomato. *Journal of Irrigation and Drainage* **32** (05), 104–106. doi:10.7631/j.issn.1672-3317.2013.05.025. (in Chinese with English abstract).

Huang, M., Zhang, Z., Zhai, Y., Lu, P. & Zhu, C. 2019 Effect of straw biochar on soil properties and wheat production under saline water irrigation. *Agronomy* **9** (8), 9080457. doi:10.3390/agronomy9080457.

Intriago, J. C., López-Gálvez, F., Allende, A., Vivaldi, G. A., Camposeo, S., Nicolás, E. N., Alarcón, J. J. & Salcedo, F. P. 2018 Agricultural reuse of municipal wastewater through an integral water reclamation management. *Journal of Environmental Management* **213**, 135–141. doi:10.1016/j.jenvman.2018.02.011.

Lei, Q. & Li, H. 2020 Effect of irrigation with secondary urban sewage on physiological characteristics of three turf grass and soil characters. *Southwest China Journal of Agricultural Sciences* **33** (11), 2545–2551. (in Chinese with English abstract).

Lesser, L. E., Mora, A., Moreau, C., Mahlknecht, J., Hernández-Antonio, A., Ramírez, A. I. & Barrios-Pina, H. 2018 Survey of 218 organic contaminants in groundwater derived from the world's largest untreated wastewater irrigation system: Mezquital Valley, Mexico. *Chemosphere* **198**, 510–521. https://doi.org/10.1016/j.chemosphere.2018.01.154.

Li, Q., Wang, X., Liu, Y., Lu, X. & Zhang, G. 2016 Effect of micro-salinity on the growth and the physiological and biochemical characteristics of sunflower. *Journal of Hebei Agricultural Sciences* **20** (01), 30–33. (in Chinese with English abstract).

Li, H., Liu, H., Pang, J., Li, S., Cui, Y. & Sun, J. 2019 Effects of water and nitrogen interaction on growth and nutrient accumulation of potted tomatoes. *Transactions of the Chinese Society for Agricultural Machinery* **50** (09), 272–279. doi:10.6041/j/issn/1000-1298.2019.09.032. (in Chinese with English abstract).

Li, D., Wan, S., Kang, Y. & Li, X. 2020 Effects of water-salt regulation on tomato growth and quality under drip irrigation with brackish water in coastal saline-alkali soil. *Journal of Irrigation and Drainage* **39** (7), 39–50. CSCD:6774540. (in Chinese with English abstract).

Liu, C., Li, Y., Guo, L., Guan, B., Liao, Y. & Wang, J. 2011 Effect of brackish water irrigation on water and salt movement in repellent soils. *Transactions of the Chinese Society of Agricultural Engineering* **27** (8), 39–45. doi:10.3969/j.issn.1002-6819.2011.08.007. (in Chinese with English abstract).

Liu, B., Wang, S., Kong, X., Liu, X. & Sun, H. 2019 Modeling and assessing feasibility of long-term brackish water irrigation in vertically homogeneous and heterogeneous cultivated lowland in the North China Plain. *Agricultural Water Management* **211**, 98–110. doi:10.1016/j.agwat.2018.09.030.

Lu, S., Zhang, X. & Liang, P. 2016 Influence of drip irrigation by reclaimed water on the dynamic change of the nitrogen element in soil and tomato yield and quality. *Journal of Cleaner Production* **139**, 561–566. doi:10.1016/j.jclepro.2016.08.013.

Morales, V. L., Parlange, J.-Y. & Steenhuis, T. S. 2010 Are preferential flow paths perpetuated by microbial activity in the soil matrix? A review. *Journal of Hydrology* **393** (1/2), 29–36. doi:10.1016/j.jhydrol.2009.12.048.

Ndour, N. Y. B., Baudoin, E., Guissé, A., Seck, M., Khouma, M. & Brauman, A. 2008 Impact of irrigation water quality on soil nitrifying and total bacterial communities. *Biology & Fertility of Soils* **44** (5), 797–803. doi:10.1007/s00374-008-0285-3.

Pan, N., Hou, Z., Chen, W., Jiao, W., Peng, C. & Liu, W. 2012 Study on soil enzyme activities and microbial biomass carbon in Greenland irrigated with reclaimed water. *Environmental Science* **33** (12), 4081–4087. doi:10.13227/j.hjkx.2012.12.004. (in Chinese with English abstract).

Pedrero, F., Camposeo, S., Pace, B., Cefola, M. & Vivaldi, G. A. 2018 Use of reclaimed wastewater on fruit quality of nectarine in Southern Italy. *Agricultural Water Management* **203**, 186–192. https://doi.org/10.1016/j.agwat.2018.01.029.

Perulli, G. D., Bresilla, K., Manfrini, L., Boini, A., Sorrenti, G., Grappadelli, L. C. & Morandi, B. 2019 Beneficial effect of secondary treated wastewater irrigation on nectarine tree physiology. *Agricultural Water Management* **221**, 120–130. https://doi.org/10.1016/j.agwat.2019.03.007.

Petousi, I., Daskalakis, G., Fountoulakis, M. S., Lydakis, D., Fletcher, L., Stentiford, E. I. & Manios, T. 2019 Effects of treated wastewater irrigation on the establishment of young grapevines. *Science of the Total Environment* **658**, 485–492. https://doi.org/10.1016/j.scitotenv.2018.12.065.

Ren, C., Zhao, Y., Gong, J., Wang, J., Li, H. & Gu, J. 2017 Spatial distribution and effect factors of soil water repellency in Guishui River Basin. *Transactions of the Chinese Society for Agricultural Machinery* **48** (10), 237–244. doi:10.6041/j.issn.1000-1298.2017.10.029. (in Chinese with English abstract).

Sanchez, W., Burgeot, T. & Porcher, J. M. 2013 A novel 'Integrated biomarker response' calculation based on reference deviation concept. *Environmental Science and Pollution Research* **20** (5), 2721–2725. doi:10.1007/s11356-012-1359-1.

Shang, Y., Li, Y. & Zhu, D. 2012 Effects of reclaimed water irrigation on soil water repellency. *Transactions of the Chinese Society OfAgricultural Engineering* **28** (21), 89–97. doi:1002-6819(2012)-21-0089-09. (in Chinese with English abstract).

Siemens, J., Huschek, G., Siebe, C. & Kaupenjohann, M. 2008 Concentrations and mobility of human pharmaceuticals in the world's largest wastewater irrigation system, Mexico City-Mezquital Valley. *Water Research* **42** (8/9), 2124–2134. https://doi.org/10.1016/j.watres.2007.11.019.

Sun, A., Cai, H., Chen, X. & Wang, J. 2007 Research on tomato growth and quality with sewage irrigation. *Journal of Irrigation and Drainage* **26** (2), 37–40. doi:10.13522/j.cnki.ggps.2007.02.009. (in Chinese with English abstract).

Tahtouh, J., Mohtar, R., Assi, A., Schwab, P., Jantrania, A., Deng, Y. & Munster, C. 2019 Impact of brackish groundwater and treated wastewater on soil chemical and mineralogical properties. *Science of the Total Environment* **647**, 99–109. doi:10.1016/j.scitotenv.2018.07.200.

Tang, S.-Q., She, D.-L. & Wang, H.-D. 2021 Effect of salinity on soil structure and soil hydraulic characteristics. *Canadian Journal of Soil Science* **101** (1), 62–73. doi:10.1139/cjss-2020-0018.

United States Salinity Laboratory Staff 1954 *Diagnosis and Improvement of Saline and Alkali Soils*. U.S. Government Printing Office, Washington DC, USA.

Viana, J. D. S., Palaretti, L. F., de Sousa, V. M., Barbosa, J. D. A., Bertino, A. M. P., de Faria, R. T. & Dalri, A. B. 2021 Saline irrigation water indices affect morphophysiological characteristics of collard. [Índices salinos da água de irrigação afetam as características morfofisiológicas da couve]. *Horticultura Brasileira* **39** (1), 79–85. doi:10.1590/s0102-0536-20210112.

Wang, L., Tian, J., Xu, G., Sheng, H. & Yan, X. 2020 The effects of drip irrigation with reclaimed wastewater on chlorophyll, photosynthesis, yield and quality of cucumber. *Journal of Irrigation and Drainage* **39** (05), 18–25. (in Chinese with English abstract).

Wu, W., Xu, C., Liu, H., Hao, Z., Ma, F. & Ma, Z. 2010 Effect of reclaimed water irrigation on yield and quality of fruity vegetables. *Transactions of the Chinese Society of Agricultural Engineering* **26** (01), 36–40. (in Chinese with English abstract).

Yan, L., Mi, F., Guo, Y., Yu, J., Liu, W. & Wang, X. 2014 Physiologic responses of Kentucky bluegrass seedling to mine recycled water stress. *Acta Agrestia Sinica* **22** (05), 1031–1037. (in Chinese with English abstract).

Yang, B., Blackwell, P. S. & Nicholson, D. F. 1994 Study on land degradation, investigation methods and improvement measures reduced by soil water repellency. *Environmental Science* **15** (4), 88–90. doi:10.13227/j.hjkx.1994.04.025. (in Chinese with English abstract).

Yang, G., Liu, S., Yan, K., Tian, L., Li, P., Li, X. & He, X. 2020a Effect of drip irrigation with brackish water on the soil chemical properties for a typical desert plant (Haloxylon Ammodendron) in the Manas River Basin. *Irrigation and Drainage* **69** (3), 460–471. doi:10.1002/ird.2419.

Yang, P., Wang, Y., Ren, S., Wei, C., He, X. & Xu, Z. 2020b Soil moisture and saline distribution characteristics and maize stem water uptake under alternate irrigation between saline water and groundwater. *Transactions of the Chinese Society for Agricultural Machinery* **51** (6), 273–281. doi:10.6041/j.issn.1000-1298.2020.06.029. (in Chinese with English abstract).

Yuan, C., Feng, S., Huo, Z. & Ji, Q. 2019 Effects of deficit irrigation with saline water on soil water-salt distribution and water use efficiency of maize for seed production in arid Northwest China. *Agricultural Water Management* **212**, 424–432. doi:10.1016/j.agwat.2018.09.019.

Zamanian, K., Pustovoytov, K. & Kuzyakov, Y. 2016 Cation exchange retards shell carbonate recrystallization: consequences for dating and paleoenvironmental reconstructions. *CATENA* **142**, 134–138. doi:10.1016/j.catena.2016.03.012.

Zhang, Y., Zhang, H., Su, Z. & Zhang, C. 2006 Effect of sewage irrigation on the distribution of heavy metal, enzyme activities and microbial population in the soil. *Journal of Safety and Environment* **6** (6), 44–50. doi:10.3969/j.issn.1009-6094.2006.06.013. (in Chinese with English abstract).

Zhang, A., Zheng, C., Li, K., Dang, H., Cao, C., Rahma, A. E., Zhang, J. & Feng, D. 2020a Responses of soil water-salt variation and cotton growth to drip irrigation with saline water in the low plain near the Bohai Sea. *Irrigation and Drainage* **69** (3), 448–459. doi:10.1002/ird.2428.

Zhang, J., Wang, Z., Zhang, J., Dou, Y. & Hou, Y. 2020b The influences of different nitrogen and salt levels interactions on fluorescence characteristics, yield and quality of processed tomato under drip irrigation. *Scientia Agricultura Sinica* **53** (5), 990–1003. CSCD:6729941. (in Chinese with English abstract).

First received 31 May 2021; accepted in revised form 2 September 2021. Available online 14 October 2021

IWA PUBLISHING

 Water Recycling and Low-Carbon Development doi: 10.2166/wrd.2021.159

Analysis on the effect of large-scale compound ecological engineering system on pollution control of the estuary of a lake

Yongbo Jiang[a], Xiaoyi Xu[a], Wei Wu[a], Bin Wang[b], Yidong Sun[a], Xiaobin Chen[a], Zhengdong Shen[a] and Tianyin Huang[a,*]

[a] School of Environmental Science and Engineering, Suzhou University of Science and Technology, Suzhou 215009, China
[b] College of Civil Engineering, Guizhou University, Guiyang 550025, China
*Corresponding author. E-mail: huangtianyin111@163.com

ABSTRACT

The quality of the water and the water environment in the estuary of a lake directly affect the water quality and ecological functions of the entire lake. Multi-technology systems, which integrate biotechnological analysis of a lake estuary and restoration of the ecological functions of the water *in situ*, have gradually been adopted for lake management and restoration. The Xielugang Estuary is located north of the Yangcheng Middle Lake and always exhibits a certain degree of eutrophic phenomena. To ensure the safety of the ecological environment in Yangcheng Lake, a multi-level purification and ecological system with 'intercept precipitation–ecological restoration–coupled biological treatment' was developed. Water quality monitoring results for the inlet and outlet of different units in the system from October 2020 to May 2021 showed that the system was effective. We also found that the purification capacity of the composite system was high and the system could significantly enhance the reduction of total nitrogen, total phosphorus, potassium permanganate index and total suspended solids. The average removal rates for these components were 41.34, 61.76, 35.21 and 67.21%, respectively, and the removal rate for typical heavy metals (Zn, Cu, Cd, and Pb) was 30.4–48.9%. The composite system substantially improved the water quality of the estuary and the wetland ecological function, demonstrating its effectiveness and significance.

Key words: composite system, ecological restoration, estuary, eutrophic phenomena, wetland ecological function

HIGHLIGHTS

- Investigated pollutant removal in large-scale estuary ecological treatment project.
- Proposed a biological treatment system coupled with *in situ* ecological engineering.
- Conducted a seasonal comparative analysis of the ecological treatment area and units.
- Provided some suggestions for the operation and management of ecological engineering.

GRAPHICAL ABSTRACT

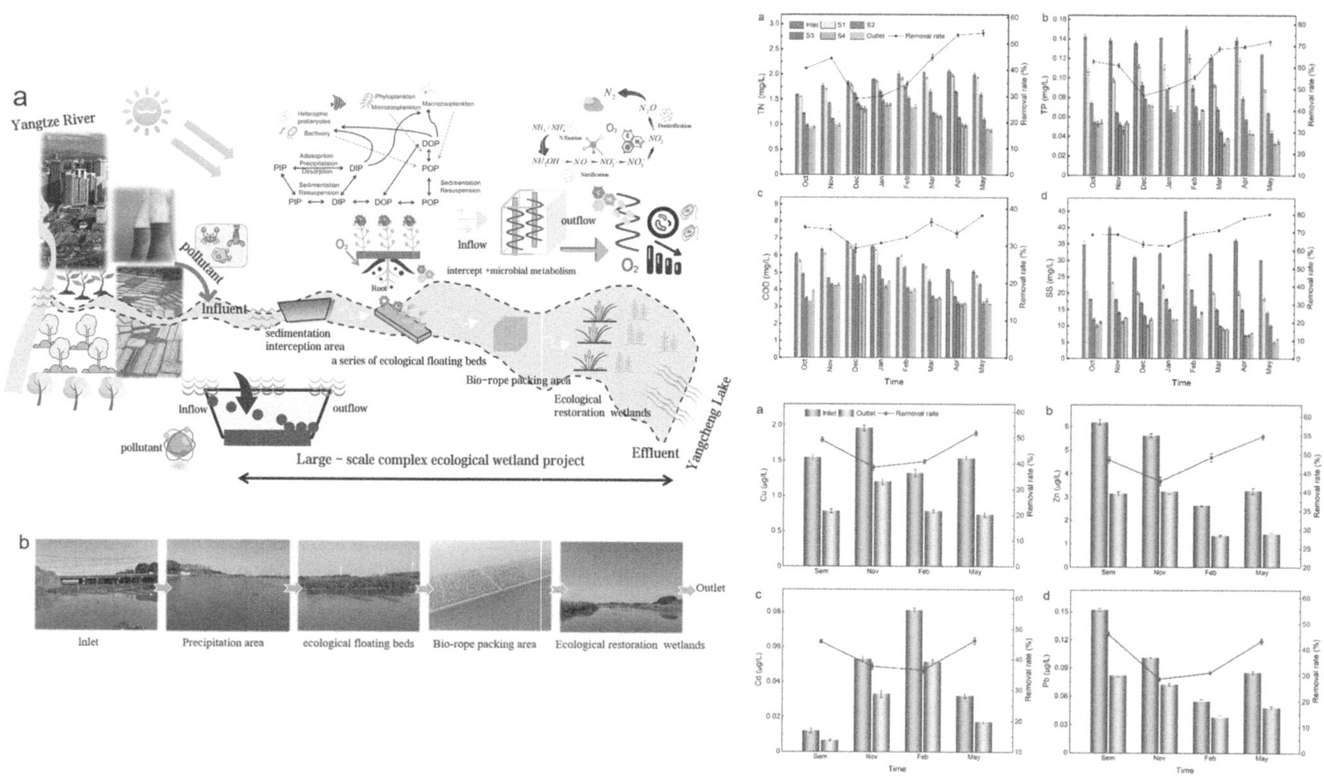

1. INTRODUCTION

Due to rapid industrialization, urbanization and population growth, the stability of aquatic ecosystems in rivers and lakes has decreased significantly worldwide, especially in municipal rivers in developing countries. To improve the water quality of lakes and rivers, it is essential for the control of pollution sources, the restructuring of river ecosystems, the regulation of water flow and the improvement of the river's self-purification (Lin *et al.* 2015; Mi *et al.* 2015). Eutrophication and algal blooms are considered the two most common and serious threats to the security of drinking water sources, nitrogen (N) and phosphorus (P) are the immediate causes of eutrophication in water, and removing them is an effective way to mitigate or prevent eutrophication (Yin *et al.* 2013; Fang *et al.* 2016). Water quality assurance and water ecological function restoration in water source areas have become important issues of common concern around the world (Nixdorff *et al.* 2021). In developing countries, due to the influence of hydraulic conditions and the topographical features of water beds, water quality pollution in estuaries has become an important reason for the destruction of lake ecosystems. Therefore, the treatment of estuary areas is not only important to reduce the pollution load that flows into lakes; it can also directly affect the self-purification function in lakes.

Currently, typical *in situ* ecological engineering techniques for water bodies include pre-storage technology, ecological floating island technology, lakefront ecological restoration technology, and artificial wetland restoration technology. However, in large-scale environmental applications, single *in situ* ecological engineering techniques often have disadvantages in terms of the purification stability and pollutant reduction (Wu *et al.* 2011, 2017; Lutterbeck *et al.* 2018; Nsenga Kumwimba *et al.* 2021). Consequently, the combination of biotechnology and engineering methods has attracted increasing attention from scholars all over the world (Sheng *et al.* 2013). At present, there are few reports on the application of this type of ecological engineering approach for river pollution control. To improve the technical applications in this field, we propose and report on a project for the ecological purification of the incoming estuary of a water body *in situ* through biotechnology-coupled engineering.

Yangcheng Lake is one of the most important freshwater lakes in the Taihu Lake Basin, and it is an important storage and ecological lake for flood control, drainage, diversion and irrigation in the Yangcheng Dianmao area, as well as an important drinking water source for Suzhou. With the increase in the population around Yangcheng Lake, the lake has taken on a large amount of pollutants while carrying drainage from upstream areas, leading to the increase of nitrogen, phosphorus, heavy metals (Zn, Cu, Cd, Pb, etc.) and other indicators entering the lake. As one of the two main waterways that enter the lake on the north shore, the Xielugang River is the main channel through which water flows into the lake from Qiputang River. This site is the object of the study in this paper. We conducted long-term monitoring of the water quality in and out of several treatment units during the water diversion period to explore the effectiveness of the estuarine ecological system for deep purification of incoming water from the upper reaches. Yangcheng Lake is an important drinking water source of Suzhou, the aim of the study was to deepen the purification of estuarine waters, reduce pollutants in the upper reaches of the estuary and provide a reference information for similar ecological restoration approaches of incoming estuaries.

2. MATERIALS AND METHODS

2.1. Study site

The study site was the compound ecological engineering system (CEES) in the Xielugang River Estuary (31°484′N–31°494′N, 120°797′E–120°816′E), situated on the northern shore of the Yangcheng Middle Lake. The water to be treated came from the Xielugang River, which feeds into the lake from Qiputang River. The Xielugang River is the boundary river between Xiangcheng City and Kunshan City in Suzhou. The main source of pollution is agricultural runoff flowing through the area. The width of the Xielugang River is about 55–65 m, and the average water depth ranges from 2.5 to 3 m (3.23 m in normal water level conditions). The flow into the lake during the diversion period ranges from 15 to 25 m^3/s, and the average river crossing section is 165.0 m^2 (All design parameters and project information from *the Design Report of Yangcheng Lake Comprehensive Remediation Project*, and same below). The Xielugang River bed meanders substantially; therefore, a pre-treatment ecological engineering area for the estuary was created to take advantage of the topography of the river. The incoming water enters the Yangcheng Middle Lake after being treated. The location and scope of the study area are shown in Figure 1.

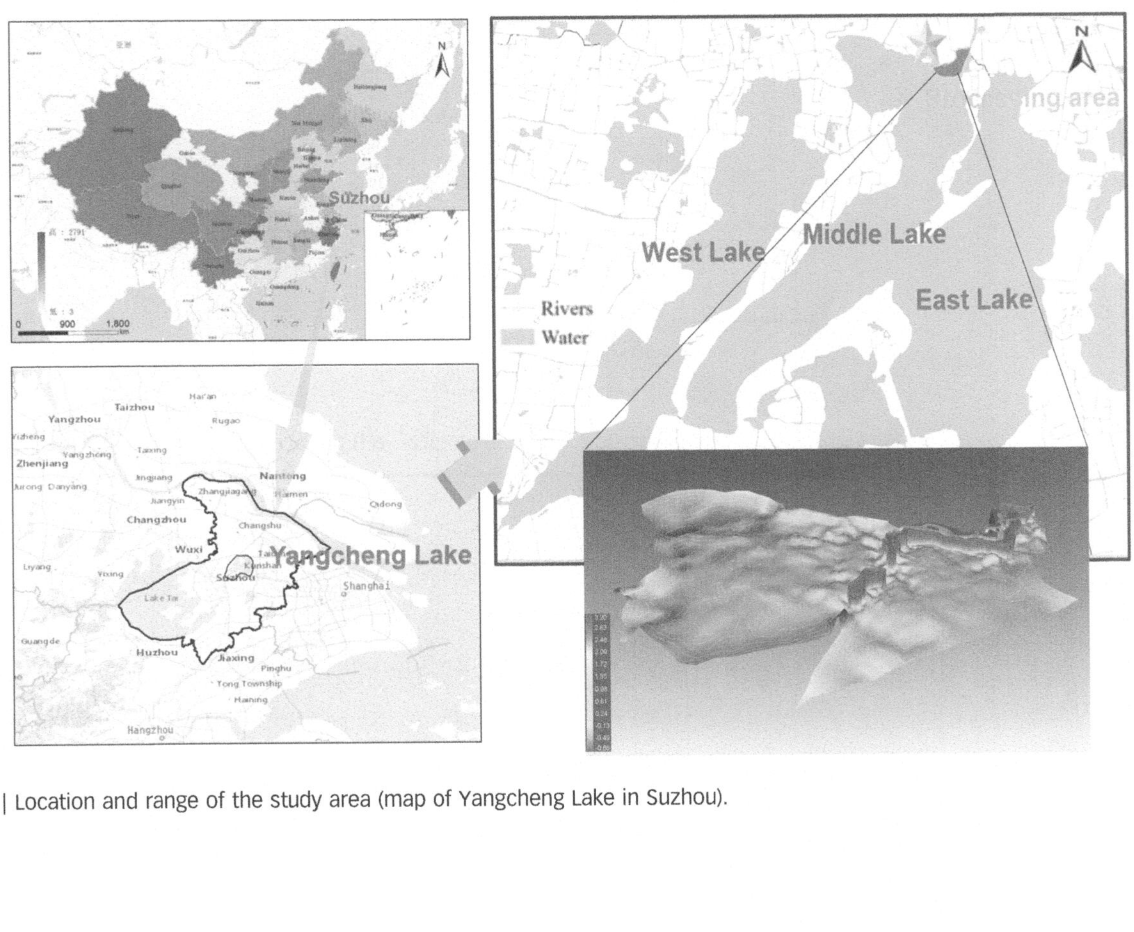

Figure 1 | Location and range of the study area (map of Yangcheng Lake in Suzhou).

2.2. Description of the composite ecological engineering process

The total area of the treatment area at the estuary of the Xielugang River is 497,900 m^2. A multi-stage ecological purification concept of ‘interception and sedimentation → strengthening purification → ecological restoration’ was applied in the treatment area, and a sedimentation interception area, ecological floating beds, biological rope filling area and ecological restoration wetlands were set up (Figure 2).

2.2.1. Sedimentation interception area

The sedimentation interception region, located in front of the estuary wetland and served as the front unit for incoming water treatment. The water body was appropriately deepened to increase the hydraulic retention time and meet the requirements for interception and sedimentation. The primary role of the sedimentation interception area was to precipitate sediment and suspended solids (SS) and to relieve the load impact of the water coming from the Xielugang River on the lake ecosystem. The engineering area has been in operation for some time and plays an important role in the initial purification of the water and reducing the impact of sediment on the subsequent wetland process. The bottom elevation of the engineering area was set at −0.5 m, the water depth was about 3.7–3.8 m under normal water level conditions and the unit covered an area of about 72,000 m^2.

2.2.2. Ecological floating beds

This unit contained a composite ecological floating bed, which was placed directly behind the sedimentation area. A total of three groups were set up in this unit with a total area of 1,944 m^2. Each group consisted of 12 rectangular modules (the frame

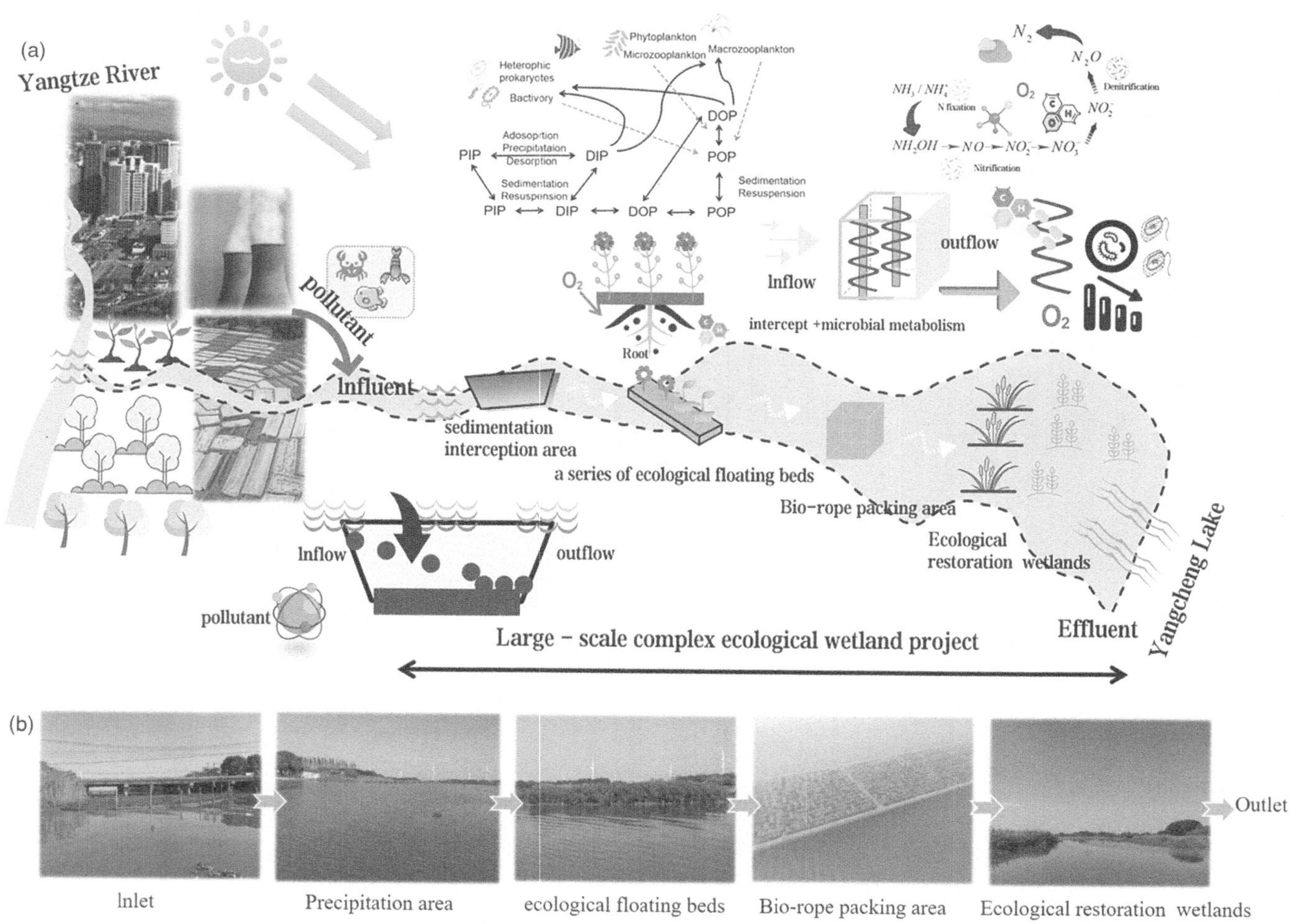

Figure 2 | The process of composite ecological engineering (a) and site photographs (b).

size of a single rectangular module was 18 × 3 × 0.5 m). The frame spacing of each rectangular module was 1.5 m, the module plane frame was welded from 40 × 50 T-type steel and 40 × 4 angle steel, and the periphery of each module was fixed to the bottom of the lake with 2–4 galvanized steel piles. The steel column was 8.3 m long and 150 mm in diameter. The pontoon (L = 3 m) was mainly made of filamentous natural plant fiber and polyester fiber with a large specific surface area and high porosity. The main plants selected for the floating beds were dry umbrella grass, *Thalia dealbata*, *Iris pseudacorus*, etc.

2.2.3. Bio-rope filling area

The biological rope filling area was laid out in a submerged stainless-steel frame totaling 1,008 m^3. The processing unit was located before the ecological wetland restoration area and was part of the bio-enhanced removal unit. The biological rope was arranged in a straight line in the main channel, with a total of four rows. To ensure maximum stability, the peripheral stainless-steel frame was welded with open hole wire. The top of the filler frame was 0.1–1.2 m below the normal water level and was completely submerged. The inlet water flowed through the filler area radially. A fiber filler with a large specific surface area, a high biological adhesion ratio and a biofilm structure was attached to the suspension rope. With different levels of oxygen, microorganisms could carry out nitrification and denitrification reactions as well as metabolism on the filler (Albuquerque *et al.* 2012).

2.2.4. Ecological restoration wetlands

The wetland restoration area in the Xielugang Estuary was a riparian surface flow wetland, which was combined with the submerged topography to create a 'deep channel water–shallow beach–land area' with different substrates. The wetland is a rich habitat with natural meandering patterns of internal water channels and roaming beaches, forming a terrain condition-rich habitat, an internal water flow channel and a floodplain with a natural winding shape. This increases the effective residence time and contact area of the wetland to meet the diverse needs of wetland species and ensure wetland purification. The total area of the wetland restoration area was about 425,900 m^2, with the deep water area and open water surface accounting for about 60% of the total area and the shallow water area accounting for 40%. The average water depth in the deep water area was maintained at 1.2 m; that is, the elevation of the lake bottom terrain was maintained at about 1.1–2 m, and the average water depth in the shallow water area was controlled within 0.3 m. In other words, the elevation of the planting area was maintained at 2.6–3.2 m, the side slope ratio was 1:5, and the slope and bottom were compacted with the original soil and covered with planting soil. To meet the needs of birds and other foraging and roosting species, a relatively shallow gravel beach was created in the shallow water area and the elevation was generally maintained at 3 m or higher. Water-holding, floating-leaf and submerged plants were planted in different functional areas. The selected water-holding plants were mainly native reeds, calamus and balsam ferns, while the submerged plants were mainly black algae, foxtail algae, hardy water lilies, Malayan eyebright and bitter grass. The ecological wetland restoration area served to clean up pollutants while fulfilling the landscape needs of the estuary and the survival needs of animals around the lake.

2.3. Field sampling and laboratory analysis

The efficacy of the CEES to minimize pollutants from the Xielugang Estuary was evaluated. From October 2020 to May 2021, we took time samples once a month, monthly water samples were collected at the inlet and outlet locations of each treatment unit in the test area for the Xielugang Estuary Compound Ecological Purification Project (precipitation trap area – S1, ecological floating beds – S2, bio-rope filling area – S3 and ecological restoration wetlands – S4). Before sampling, all sampling equipment and storage containers were cleaned with distilled water. During field sampling, three parallel water samples were collected at 30 cm below each sampling point without disturbing the sediment–water interface and mixed in 500 mL polyethylene bottles. The water samples were labeled and sealed with airtight screw caps. All collected water samples were immediately refrigerated and transported on the day of sampling to the laboratory, where they were stored in a low temperature environment and subjected to analytical experiments within 48 h.

The total suspended solids (TSS), total nitrogen (TN), chemical oxygen demand (COD_{Mn}) and total dissolved phosphorus (TP) were measured using the standard procedures specified in the Analytical Methods for Water and Wastewater Monitoring (4th edition, 2002). Dissolved oxygen (DO), pH and temperature were measured using the YSI 550A Handheld Dissolved Oxygen and Temperature System purchased from TechTrend International Limited, USA. A Secchi disc was used to measure water clarity. In the laboratory, TN was directly analyzed with alkaline potassium persulfate digestion-ultraviolet-visible (UV) spectrophotometry in unfiltered water samples, and TP was directly analyzed using the ammonium molybdate spectrophotometric method. TSS was calculated by dry weighing analysis and COD_{Mn} was analyzed with HACH ultra-low range ablation spectrophotometry.

The seasonal removal of heavy metals (Cu, Zn, Pb, Cd, etc.) by the whole system in the test area was also evaluated. In September 2020, November 2020, February 2021 and May 2021, samples were collected at the inlet and outlet of the estuary test area. Water samples were pre-filtered through a 0.45 um microporous membrane, subjected to microwave digestion (CEM Mars6, USA) and then analyzed with inductively coupled plasma mass spectrometry (ICP-MS, Agilent 7700, USA). All analyses were carried out according to the standards defined in the Analytical Methods for Water and Wastewater Monitoring.

3. RESULTS AND DISCUSSION

The monthly average concentrations of different nutrients and conventional physical and chemical indicators before and after treatment with the CEES (sampling and analysis times were from October 2020 to May 2021) are listed in Table 1. The average monthly effluent from the treatment area fully met the Class IV standard in the Environmental Quality Standard for Surface Water (GB 3838-2002), and some indexes involved met the requirement of the Class III level.

3.1. The removal rate of TN

Nitrogen is one of the main factors that lead to eutrophication in lake and reservoir water bodies. The primary methods through which the ecological purification system removes nitrogen from water bodies are nitrification and denitrification reactions by microorganisms, absorption and transformation by plants and adsorption by filler media (Vymazal 2007; Nsenga Kumwimba *et al.* 2021). The changes in concentration at different sampling locations at different times during the test period are shown in Figure 3(a). The removal rate of TN by the ecological purification system in the estuary was also calculated.

As can be seen from Figure 3(a), the sedimentation interception area had a low contribution to TN removal. The purification and removal of TN relied on synergistic effects from the ecological floating bed zone, the biological rope filling zone and the wetland restoration zone. The removal rate of TN in the whole ecological purification system ranged from 29 to 55%. The removal rate in May was the highest and was 1.9 times higher than that in December. The biological rope filler area played a leading role in TN removal, and the highest contribution from this area reached 43.13%. The removal rates in autumn and spring were obviously higher than that in winter, mainly because a large number of plants in the floating bed and wetland withered, died and were not harvested in the winter, which had a substantial impact on TN absorption, absorption and transformation. The low temperatures also influence the metabolism of microorganism, and this reduces the effectiveness of the system to treat these pollutants in the winter (Newcomer Johnson *et al.* 2016). In addition, the nitrifying and denitrifying microorganisms on the biological rope filler were less active in low temperature conditions (Sheng *et al.* 2013). As the temperature rose, the removal rate increased, reaching the highest value in May. The overall operation of the ecological purification system was stable and the impact of this area was more obvious in the removal of TN from incoming water from the upper reaches of the lake. In addition, seasonal variations had some impact on the operation of the system in the project area (Vymazal 2007; Li *et al.* 2020).

Table 1 | The parameters for river water sample (unit: mg L^{-1})

Item	Sort	TN	TP	TSS	COD_{Mn}	DO	pH	Water clarity (cm)
Inlet	Average	1.90	0.136	35.25	6.05	7.45	7.98	48
	Range	1.60–2.06	0.12–0.15	30–45	4.8–7.2	6.78–8.96	7.50–8.43	–
Outlet	Average	1.08	0.052	11.56	3.92	8.24	8.08	73
	Range	0.88–1.32	0.04–0.07	6.0–12.5	3.15–4.3	7.02–9.23	7.46–8.43	–
Standard (IV)	–	1.50	0.10	–	10	≥3	6–9	–
Standard (III)	–	1.00	0.05	–	6	≥5	6–9	–
Removal rate (%)	–	28.96–54.17	47.02–71.23	62.50–80.12	29.41–36.36	–	–	–

These values are means calculated from the data from October 2020 to May 2021.

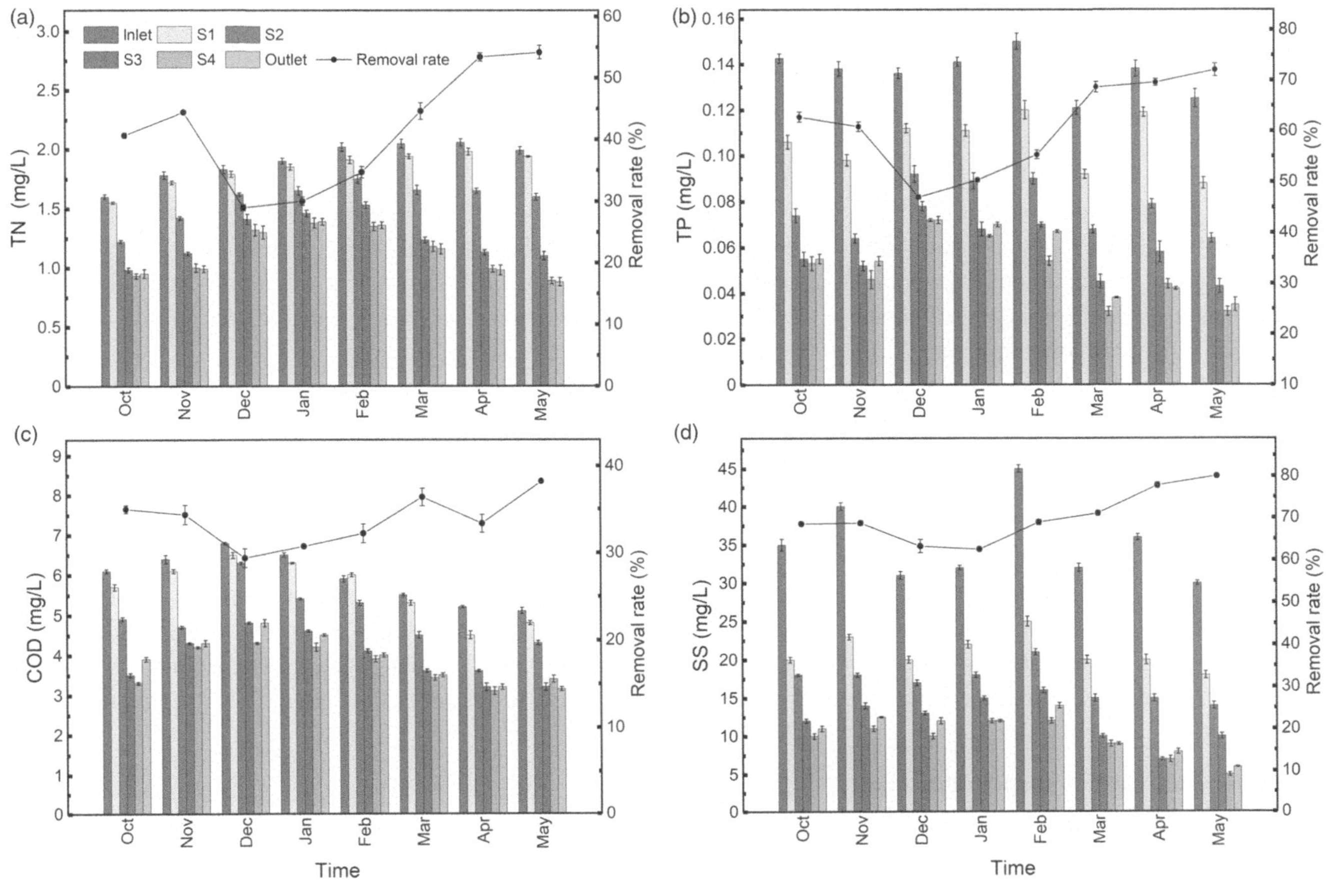

Figure 3 | Variations and removal rates of TN (a), TP (b), COD (c) and SS (d) for the system.

3.2. The removal rate of TP

The phosphorus removal methods in the ecological purification system were different from those used for nitrogen removal. The removal methods for phosphorus were complex and involved physical, biological and chemical effects. The physical effects mainly consisted of pollutants gradually settling at the bottom of the precipitation area due to gravity and interception by emergent plants in shoals (Nsenga Kumwimba *et al.* 2021). The biological action mainly consisted of microbial uptake, plant and algae uptake at the biological rope filler and plant roots. Plants and algae need to absorb a large amount of soluble phosphate for growth, and microorganisms need to absorb a large amount of inorganic phosphorus, which is assimilated into their bodies in the form of polyphosphate and fixed in the process of growth. (Vymazal 2007; Wu *et al.* 2017; Nsenga Kumwimba *et al.* 2021). The higher the DO is, the more phosphorus will be absorbed. The chemical effects mainly consisted of phosphorus binding to ions in the matrix filler to form particles that were intercepted and precipitated for removal and adsorption on the surface of the matrix (Benvenuti *et al.* 2018). Elemental phosphorus is a major factor in the eutrophication of lake and reservoir waters. Variations in the TP concentration at different sampling locations at different times during the experiment are shown in Figure 3(b). The removal rate of TP by the estuarine ecological purification system was calculated for matrix surface adsorption.

Figure 3(b) shows that the removal rate of TP in the precipitation interception area was greater than that of TN. The average contribution of this area in TP removal was maintained at about 37.03%, primarily due to the sedimentation from the particulate phosphorus in TP. The waterfront plants on both sides of the shoreline also played a part in the interception and sedimentation effect.

The TP removal rate of the whole ecological purification system ranged from 47 to 72%. The highest removal rate was observed in May and was as high as 72%. The removal rate in winter was relatively low, because the plants growth and microbial metabolism required less nutrients accordingly under the conditions of low temperature. The microorganism metabolism was also slower and less active at low temperatures in winter. Excess phosphorus originally absorbed by the polyphosphorus bacteria may also be released in low DO conditions (Newcomer Johnson *et al.* 2016). Therefore, TP removal in the winter mainly relies on physical sedimentation, adsorption by substrate fillers and interception by the roots of water-holding plants, submerged plants, etc. To improve the phosphorus removal capacity of the system, it is necessary to improve the interception settlement capacity (Lin *et al.* 2015; Mi *et al.* 2015).

3.3. The removal rate of COD_{Mn}

COD_{Mn} is one of the indicators of the relative content of organic matter in water. The removal of COD_{Mn} from water bodies by ecological purification systems mainly occurs through interception and adsorption by substrates, absorption by plants, algal root growth, precipitation and microbial degradation. Related studies have shown that microbial degradation is the most important method (Newcomer Johnson *et al.* 2016; Nsenga Kumwimba *et al.* 2021). Specific concentration changes and removal rates are shown in Figure 3(c).

Figure 3(c) shows that the removal rate of COD_{Mn} for the whole ecological purification system was weaker than those of nitrogen and phosphorus, and the overall removal rate ranged from 30 to 38%. Comparison of each treatment unit revealed that the degradation rate of COD_{Mn} in the bio-rope packing area was significantly greater than that in the other unit treatment areas, with an average contribution of 44.36%. The biological rope packing area mainly relied on the metabolic degradation of a large number of microorganisms attached to inorganic substances on the surface layer of the filler. Insoluble organic matter was filtered and intercepted by the substrate in the wetland system, and some dissolved organic matter was directly absorbed by plants and algae to synthesize their own substances (Fang *et al.* 2016). The roots of plants also absorbed and adsorbed a certain amount of dissolved organic matter, which was then removed by the action of biofilm on their surface. Therefore, wetland plants in the ecosystem should be harvested in time to prevent residual organic matter from being retained and released, to ensure the treatment effectiveness and water safety in the project area (Sheng *et al.* 2013; Yin *et al.* 2013; Fang *et al.* 2016).

3.4. The removal rate of suspended solids

The main methods by which the ecological purification system removed suspended solids in the water body were natural gravity sedimentation, plant root adsorption, interception, etc. (Vymazal 2007; Sheng *et al.* 2013).

As can be seen from Figure 3(d), the whole ecological purification system had a high rate of SS removal. The removal efficiency ranged from 61 to 80%, which effectively increased the transparency of the water body. The sedimentation interception area had the most significant effect on SS removal, with a monthly average contribution of 44.4%, mainly because the sedimentation area had a slower flow rate, which maximized the natural sedimentation effect of SS. The removal rate in winter was slightly lower than that in other seasons, further indicating that wetland plants have a strong interception effect on SS (Fang *et al.* 2016). In addition, the removal rates of SS and TP showed a positive correlation, indicating that greater interception and sedimentation of SS, which is a natural substrate carrier and adsorbent, was related to better TP removal (Sheng *et al.* 2013; Nsenga Kumwimba *et al.* 2021).

3.5. Trace metal reduction efficiency

The removal mechanism for heavy metals in the estuary complex ecological purification system mainly consisted of matrix adsorption, filtration, precipitation, plant absorption and microbial degradation. Many studies have shown that in wetland treatment systems, clay particles and organic matter on the surface of the substrates play a major role in the adsorption and precipitation of heavy metals, and that absorption and root adsorption by plants have little effect on the removal of heavy metals (Fang *et al.* 2016; Colares *et al.* 2020). The migration and transformation of heavy metals between the substrate and the water body refers to the adsorption and release of heavy metals by solid substrate particles, as well as the physical, chemical, and biological processes that accompany this process (Arini *et al.* 2012; Colares *et al.* 2020). The main physical processes were adsorption, filtration and deposition. The chemical processes involved mainly included adsorption, precipitation, ion exchange, oxidation and hydrolysis. The substrate in ecological wetlands is an active filter. After the river water enters the ecological wetland system, the suspended matter is intercepted, filtered and deposited in the matrix. The heavy metals in the water body interact with the surface clay particles in the matrix for the adsorption and enrichment of heavy metals. Therefore, the surface of fillers is the location where heavy metals gather, and it is also a potential source of heavy metal pollution. Proper disposal of the surface matrix is particularly critical. The migration, transformation and accumulation of heavy metals in plants mainly occur due to plants directly absorbing, adsorbing and enriching heavy metals from polluted water bodies (Rezania *et al.* 2016). Heavy metals migrate from the roots to the above-ground parts in

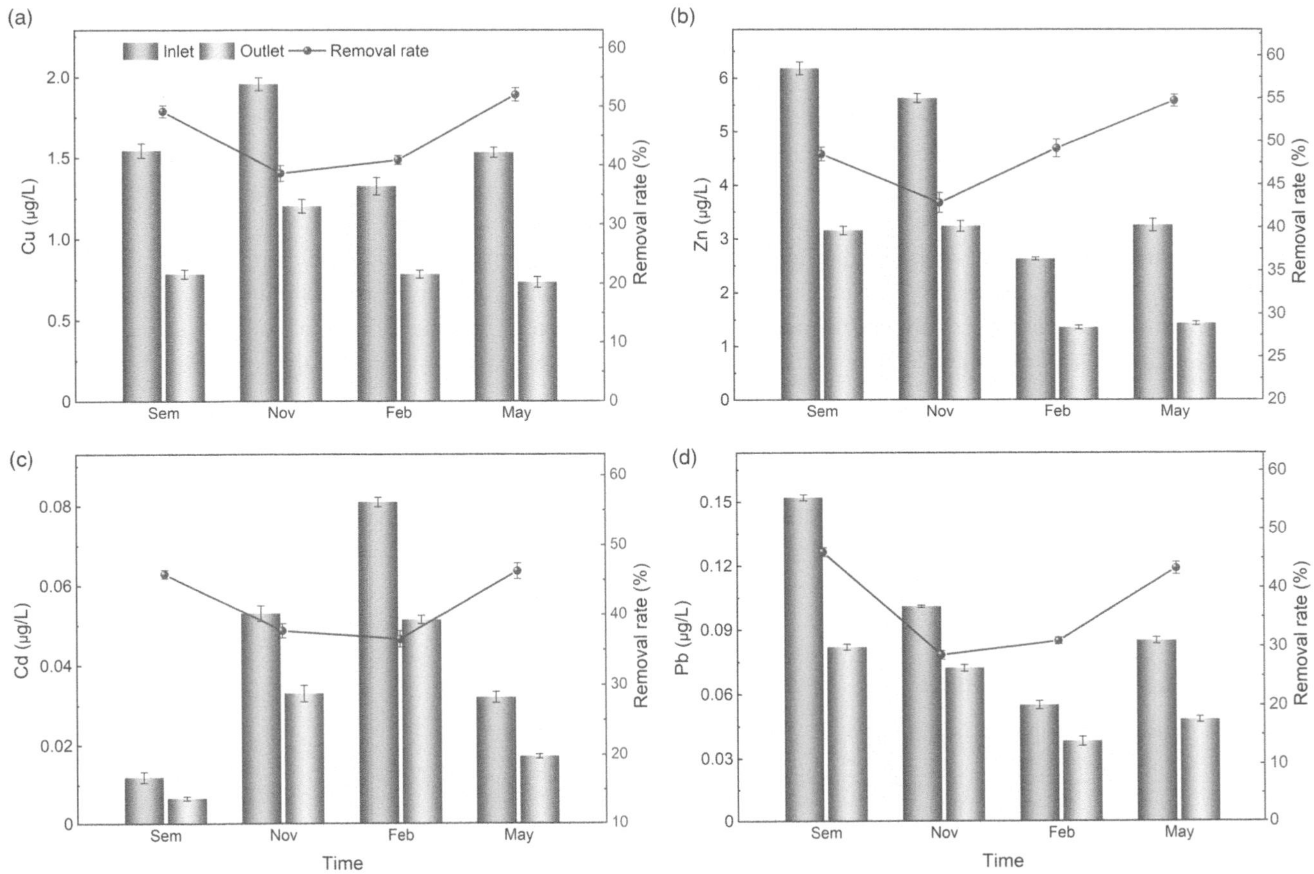

Figure 4 | Variations and removal rates of Cu (a), Zn (b), Cd (c) and Pb (d) for the system.

plants and are removed when the above-ground parts are harvested. The activity of heavy metals is reduced by plants. To facilitate the precipitation of heavy metals and inhibit their migration, wetland plants should be harvested regularly in autumn and winter to avoid potential risks (Gill *et al.* 2017).

Figure 4 shows that the average removal rate of heavy metals by the composite ecological purification system was higher in September (47.52%) and May (49.41%) than in November (36.83%) and February (39.3%). In terms of the average removal rate of individual heavy metals, Zn (48.9%) > Cu (45.1%) > Cd (41.6%) > Pb (30.4%). The seasonal changes of pollutants removal were obvious, mainly because the plants in the estuary purification system had stopped growing and adsorbing, which resulted in a poor removal rate for heavy metals. The amount of water drawn into the lake from Qiputang River decreased in the winter, and the flow rate of the water was slow. The removal of heavy metals mainly depended on natural precipitation. In summer, the temperature was high, the plants in the system grew the most luxuriant and the developed root tissues of the plants enhanced the absorption and transport of heavy metal ions (Rezania *et al.* 2016; Wu *et al.* 2017). Though the heavy metal content in the influent water in summer was high, CEES maintains higher removal rates. The removal rate of heavy metal ions in constructed wetlands was highest in the summer (Colares *et al.* 2020).

4. CONCLUSION

1. During the operation period for the estuary coupled with a purification system, the treatment effect of each unit was relatively stable, and the content of typical pollutant entering the lake met the Class IV standard of *Surface Water Environmental Quality Standard* (GB 3838-2002). The average removal rates for TP, COD_{Mn} and SS were 41.34, 60.82, 33.75 and 69.85%, and the average removal rates for heavy metals Zn, Cu, Cd and Pb in water were 48.9, 45.1, 41.6 and 30.4%, respectively. Seasonal changes had a greater impact on the overall operation of the system, and the removal rate for the entire system in winter was not as good as that in autumn and spring.

2. With regard to the characteristics of the removal of different pollutants, each treatment unit in the coupling system had individual merits. The sedimentation zone had a good purification effect on TP and SS, and was less affected by seasonality. The bio-rope packing area played a leading role in the removal of TN and COD_{Mn}, which are greatly affected by seasonality. The removal efficiencies of different pollutants in the ecological floating bed area and the wetland restoration area were relatively stable, and the ecological restoration function was more apparent. Therefore, the units of the estuary complex ecological system played an effective complementary role in time and space, and substantially improved the estuary water quality and wetland water ecological function.
3. In system management, it is necessary to pay attention to the harvesting and disposal of plants in the floating island area and the wetland restoration area. It is also necessary to pay close attention to the seasonal filming of the biological rope filling area and regularly collect samples for microbial character testing. In terms of internal sources, the sedimentation interception area has a long running time and regular dredging and remediation work should be carried out.

ACKNOWLEDGEMENTS

This study was financially supported by the Suzhou Science and Technology Plan Project – Minsheng Science and Technology (Grant No. SS202002) and the Water Pollution Control and Treatment National Science and Technology Major Project (Grant No. 2017ZX07205002).

DATA AVAILABILITY STATEMENT

All relevant data are included in the paper or its Supplementary Information.

REFERENCES

Albuquerque, A., Makinia, J. & Pagilla, K. 2012 Impact of aeration conditions on the removal of low concentrations of nitrogen in a tertiary partially aerated biological filter. *Ecological Engineering* **44**, 44–52. https://doi.org/10.1016/j.ecoleng.2012.03.006.

Arini, A., Feurtet-Mazel, A., Morin, S., Maury-Brachet, R., Coste, M. & Delmas, F. 2012 Remediation of a watershed contaminated by heavy metals: a 2-year field biomonitoring of periphytic biofilms. *Science of the Total Environment* **425**, 242–253. https://doi.org/10.1016/j.scitotenv.2012.02.067.

Benvenuti, T., Hamerski, F., Giacobbo, A., Bernardes, A. M., Zoppas-Ferreira, J. & Rodrigues, M. A. S. 2018 Constructed floating wetland for the treatment of domestic sewage: a real-scale study. *Journal of Environmental Chemical Engineering* **6** (5), 5706–5711. https://doi.org/10.1016/j.jece.2018.08.067.

Colares, G. S., Dell'Osbel, N., Wiesel, P. G., Oliveira, G. A., Lemos, P. H. Z., Da Silva, F. P., Lutterbeck, C. A., Kist, L. T. & Machado, Ê. L. 2020 Floating treatment wetlands: a review and bibliometric analysis. *Science of the Total Environment* **714**, 136776. https://doi.org/10.1016/j.scitotenv.2020.136776.

Fang, T., Bao, S., Sima, X., Jiang, H., Zhu, W. & Tang, W. 2016 Study on the application of integrated eco-engineering in purifying eutrophic river waters. *Ecological Engineering* **94**, 320–328. https://doi.org/10.1016/j.ecoleng.2016.06.003.

Gill, L. W., Ring, P., Casey, B., Higgins, N. M. P. & Johnston, P. M. 2017 Long term heavy metal removal by a constructed wetland treating rainfall runoff from a motorway. *Science of the Total Environment* **601–602**, 32–44. https://doi.org/10.1016/j.scitotenv.2017.05.182.

Li, Y., Zhang, H., Zhu, L., Chen, H., Du, G., Gao, X. & Pu, Y. 2020 Evaluation of the long-term performance in a large-scale integrated surface flow constructed wetland–pond system: a case study. *Bioresource Technology* **309**, 123310. https://doi.org/10.1016/j.biortech.2020.123310.

Lin, M., Li, Z., Liu, J., Gozlan, R. E., Sovan, L., Zhang, T., Ye, S., Wei, L., Jing, Y. & Robert, B. 2015 Maintaining economic value of ecosystem services whilst reducing environmental cost: a way to achieve freshwater restoration in China. *PLoS One* **10** (3), e120298.

Lutterbeck, C. A., Zerwes, F. V., Radtke, J. F., Köhler, A., Kist, L. T. & Machado, Ê. L. 2018 Integrated system with constructed wetlands for the treatment of domestic wastewaters generated at a rural property – evaluation of general parameters ecotoxicity and cytogenetics. *Ecological Engineering* **115**, 1–8. https://doi.org/10.1016/j.ecoleng.2018.01.004.

Mi, Y., He, C., Bian, H., Cai, Y., Sheng, L. & Ma, L. 2015 Ecological engineering restoration of a non-point source polluted river in Northern China. *Ecological Engineering* **76**, 142–150. https://doi.org/10.1016/j.ecoleng.2014.05.004.

Newcomer Johnson, T. A., Kaushal, S. S., Mayer, P. M., Smith, R. M. & Sivirichi, G. M. 2016 Nutrient retention in restored streams and rivers: a global review and synthesis. *Water* **8** (4), 116.

Nixdorff, H., Noga, J., Amsalu, D., Springett, J. & Ashbolt, N. 2021 Improving the implementation of water and resource recovery in Canada. *Journal of Water Reuse and Desalination*. doi:10.2166/wrd.2021.087.

Nsenga Kumwimba, M., Li, X., Wang, W., De Silva, L. H. D. K., Bao, L., Mihiranga, H. K. M., Su, J. & Li, X. 2021 Large-scale hybrid accidental urban wetland for polluted river purification in northern China: evidence and implications for urban river management. *Environmental Technology & Innovation* **22**, 101542. https://doi.org/10.1016/j.eti.2021.101542.

Rezania, S., Taib, S. M., Md Din, M. F., Dahalan, F. A. & Kamyab, H. 2016 Comprehensive review on phytotechnology: heavy metals removal by diverse aquatic plants species from wastewater. *Journal of Hazardous Materials* **318**, 587–599. https://doi.org/10.1016/j.jhazmat.2016.07.053.

Sheng, Y., Qu, Y., Ding, C., Sun, Q. & Mortimer, R. J. G. 2013 A combined application of different engineering and biological techniques to remediate a heavily polluted river. *Ecological Engineering* **57**, 1–7. https://doi.org/10.1016/j.ecoleng.2013.04.004.

Vymazal, J. 2007 Removal of nutrients in various types of constructed wetlands. *Science of the Total Environment* **380** (1), 48–65. https://doi.org/10.1016/j.scitotenv.2006.09.014.

Wu, H., Zhang, J., Li, P., Zhang, J., Xie, H. & Zhang, B. 2011 Nutrient removal in constructed microcosm wetlands for treating polluted river water in northern China. *Ecological Engineering* **37** (4), 560–568. https://doi.org/10.1016/j.ecoleng.2010.11.020.

Wu, Y., Han, R., Yang, X., Zhang, Y. & Zhang, R. 2017 Long-term performance of an integrated constructed wetland for advanced treatment of mixed wastewater. *Ecological Engineering* **99**, 91–98. https://doi.org/10.1016/j.ecoleng.2016.11.032.

Yin, H., Kong, M. & Fan, C. 2013 Batch investigations on P immobilization from wastewaters and sediment using natural calcium rich sepiolite as a reactive material. *Water Research* **47** (13), 4247–4258. https://doi.org/10.1016/j.watres.2013.04.044.

First received 14 June 2021; accepted in revised form 30 September 2021. Available online 21 October 2021